DEVELOPMENT BUILDING: THE TEAM APPROACH

THE AMERICAN
INSTITUTE OF
ARCHITECTS
WASHINGTON, D. C.

DEVELOPMENT BUILDING: THE TEAM APPROACH

By C. W. Griffin

Illustrated by
Charles D. Stokes

Distributed by Halsted Press Division,
John Wiley & Sons, Inc., New York, N. Y.

This book was set in Optima with Helvetica heads by Hodges Typographers Incorporated. The book was designed, illustrated, edited and produced by the AIA Publishing Department, Wm. Dudley Hunt, Jr., FAIA, Publishing Director; Fredrick H. Goldbecker, Manager, Publishing Services; Victoria Wakefield, Editor; Charles D. Stokes, Designer.

Quotations from William Zeckendorf reprinted from and based upon Zeckendorf, by William Zeckendorf with Edward McCreary. Copyright © 1970 by William Zeckendorf. Reprinted by permission of Holt, Rinehart and Winston, Inc.

Quotation by William Caudill reprinted from Architecture By Team: A New Concept For The Practice of Architecture, by William Wayne Caudill, FAIA. Copyright © 1971. Reprinted by permission of Van Nostrand Reinhold.

First Printing, May 1972
Second Printing, April 1973
ISBN   0-913962-00-7

Distributed by Halsted Press Division,
John Wiley & Sons, Inc.. New York, N.Y.

Library of Congress Catalog Card Number
72-79441

M 135

PREFACE

"I early found that when I worked for myself alone, myself alone worked for me; but when I worked for others also, others worked for me."
— Benjamin Franklin

The American Institute of Architect's purpose in publishing this book is twofold: To alert architects and other design professionals to a promising new opportunity for expanding the scope of architectural practice as members of project development teams; to explain the mechanics of the development process itself.

Before now the information was scattered throughout a variety of sources —real estate, banking, engineering, and architectural periodicals. This represents the first effort to collate the information and disseminate it.

The book is intended to be a large-scale road map, guiding the architect toward his goal of successful project development, alerting him to the dangers of profitless and hazardous detours. The development process is too complex and varied to be defined in simple, unvarying formulas. What this book attempts to do is to prescribe general rules. When we say, for example, that the mortgage banker negotiates the mortgage loan commitment, we do not imply that this always is the case. An architect might negotiate a mortgage loan in a country club locker room. When we advise an architect not to invest his entire fee as equity share in a development project, we realize that some architects may have good personal reasons for doing otherwise.

The primary audience comprises architects. But the book should appeal also to experts from other disciplines involved in the development process. Among these are engineers, lawyers, contractors, mortgage bankers, corporate owners, real estate brokers, and public officials.

The book grew out of informal studies by the AIA's national Committee on Professional Consultants. Late in 1968, the committee began investigating new trends in architectural practice and began to identify groups playing key roles in the early decision-making stage of the building process. The group met with financiers, real estate experts, lawyers, contractors, public officials, and other development decision makers. Many professional and business associations were represented at these sessions—the National Society of Professional Engineers, the Consulting Engineers Council of the U.S., the American Society of Real Estate Counselors, the Associated General Contractors of America, the Mortgage Bankers Association of America, the National Association of Real Estate Boards, the U.S. Savings and Loan League, and the Urban Land Institute.

In 1971, to focus continued attention on the subject, a Task Force on The Architect in the Development Team was created. The four-man task force comprised Chairman Herbert E. Duncan, Jr., AIA, of Kansas City, Mo.; Harry A. Golemon, AIA, of Houston; Michael Maas, AIA, of New York; and Robert L. Stelzl, of Los Angeles. Carl L. Bradley, AIA, of Fort Wayne, Ind., served as liaison with the AIA Board of Directors. Robert Allan Class, AIA, provided AIA staff liaison.

Thus was born the present volume.

ACKNOWLEDGEMENTS

This book was written with the help of many individuals who contributed special knowledge of the development process.

Preston M. Bolton, FAIA, secretary
The American Institute of Architects

Gil P. Bourk, president
First Mortgage Investment Co.
Kansas City, Mo.

Howard B. Christy, Jr.,
    regional planner
The Houston-Galveston Area Council

J. Sprigg Duvall, president
Victor O. Schinnerer & Co., Inc.
Insurance Brokers
Washington, D. C.

Dale Elmore
New Towne Investment
    Properties, Ltd.
Beaumont, Texas

Robert W. Jones, AIA, vice president
Shreve, Lamb & Harmon Associates
New York City

Russell S. Jones, vice president
Herbert V. Jones & Co., Realtors
Kansas City, Mo.

Harry Lane
Century Development Corp.
Houston, Texas

Harry T. Maulding, president
Mortgage Bankers Association of
    Metropolitan Washington
Washington, D. C.

Robert F. McConkey,
    senior vice president
Perpetual Building Association
Washington, D. C.

Howard N. Neilson,
    second vice president
Mortgage and Real Estate
    Department
Connecticut General
    Life Insurance Co.
Hartford, Conn.

John M. Parker, vice president
The Corporate Appraisal Co.
Cincinnati, O., and New York City

Mrs. Judith E. R. Roeder, associate
Deeter Richey Sippel Associates
Pittsburgh, Pa.

Frederick P. Rose, president
Rose Associates
New York City

L. Arnold Schafer,
    senior vice president
Weaver Bros. Realtors and
    Mortgage Bankers
Washington, D. C.

Ronald S. Senseman, FAIA, president
Ronald S. Senseman, Architects
Langley Park, Md.

Shindler Cummins, Inc.
Investment Realtors
Houston, Texas

James Shivler, Jr., P.E., president
National Society of
    Professional Engineers

Douglas Simpkins, Jr.
Ben G. McGuire & Co.
Mortgage Bankers
Houston, Texas

Jack Sonnenblick,
    executive vice president
Sonnenblick-Goldman Corp.
Mortgage Bankers & Realtors
New York City

James E. Stevens,
    assistant vice president
Victor O. Schinnerer & Co., Inc.
Insurance Brokers
Washington, D. C.

Mrs. Marjorie Weber, office manager
Sonnenblick-Goldman Corp.
Mortgage Bankers & Realtors
New York City

Harry Wenning, AIA
Wenning & D'Angelo Associates
Hastings-on-Hudson, N. Y.

John N. Wetmore, research director
Mortgage Bankers Association of
    America
Washington, D. C.

Frank J. Whalen, Jr.
Spencer, Whalen & Graham
Attorneys
Washington, D. C.

The expertise provided by AIA staff members Arthur T. Kornblut, AIA, J. Winfield Rankin, Hon. AIA, and William Wolverton, in reviewing the manuscript and particularly the chapters on liability, ethics and financing, is appreciated by the author and the task force members.

FOREWORD

Pioneering design professionals all over the United States are expanding architectural practice into a new domain — project development. Project development is a step beyond the traditional architectural role — design. In their traditional roles, design professionals not always have been permitted to maintain control over design. Too often, they have been retained only after crucial decisions have been made — decisions which significantly affect design. This book charts the course for architects and other design professionals to take a leadership position in development building and to exercise control over design quality. The author has drawn from the expertise of architects, realtors, mortgage bankers, engineers, lawyers — decision makers — to produce a definitive book on new opportunities.

Max O. Urbahn, FAIA
President
The American Institute of
   Architects

CONTENTS

I

A NEW DIMENSION IN ARCHITECTURAL PRACTICE

"Once I asked Julian Bond, of Bond Stores,
'Do you give your architects much leeway in the design of stores?'
  He snapped back, 'Does your secretary dictate your letters?' "
                                          — William Zeckendorf

Pioneering architects all over the United States are expanding architectural practice into a new domain—project development. Traditionally restricted to the central part of the three-stage decision-design-delivery process, architects are breaking out of their old roles and taking on new responsibilities at both ends. They are entering the decision stage, in some instances as project co-owners, in others as consultants offering clients new services in programming and economic analysis. As construction managers, architects are extending their services in another direction, offering cost control through the design stage and construction scheduling and coordination during the delivery stage.

The scope of these architect-managed projects is tremendous. Large architectural firms are taking a piece of the action in huge hotel-office-retail complexes and even new towns. At the opposite end of the project scale, small architectural firms are doing the same in office buildings for three or four tenants. Between the two poles is a full spectrum of project sizes and types—industrial buildings, motels, highrise office buildings, garden and highrise apartments, shopping centers, nursing homes, medical centers, and even hospitals—all with architects playing an entrepreneurial as well as a designer's role. Even public projects are involved, with architects doubling as joint-venture developers in turnkey public housing projects—projects designed and built by private developers and then sold to a public housing authority.

THE IMPORTANCE OF BEING EARLY   Why should architects want to expand their services into an entrepreneurial role? What's wrong with maintaining their traditional design and design-related ancillary services? Why should architects want to participate in the owner's decisions?

Better design control is the unanimous reply from architects who have strengthened their role in the decision stage. Programming and budgeting decisions made in the decision stage without the architect's participation can strip him of effective design control. Excluding the architect from the program-writing phase of the decision stage may lock the design into a preconceived solution, severely limiting the architect's design options. This restriction holds even when the project budget is based on accurate cost estimates. When the architect makes a late entry into the development process, the reduced time allowed for completion of preliminary and final design intensifies the pressure for a fast design solution. Lacking time to study imaginative alternatives, the architect is more likely to stand by an old tried and tested, but possibly inferior, design concept (see page 3).

When the budget is set too low, the problem is obviously far worse. A program tied to an unrealistic budget can cost an architect lost fees, additional design work, and damage to his professional reputation. It puts the architect in the position of a doctor whose diabetic patient is on a self-prescribed, high-starch diet. An unreasonable program-budget combination does not present a challenge that the architect should attempt to meet. Indeed, it is an archaic and inefficient practice. Participation in the decision stage is essential for an architect to fulfill his professional obligations. Architectural design is growing ever more dependent on analyses made during the decision stage, as indicated in the AIA report, Creating the Human Environment:

". . . Maximizing the architect's contributions before too many options have been foreclosed should be a major role of the future . . . technical aids for decision-making and for conceptual and schematic design may have far greater influence than myriad programs aimed at more efficient handling of technical details."

As a co-owning, equity-sharing partner on the development team, the architect can demand a key role in the decision stage. Even as a non-owner consultant, he can contribute vital technical data and provide creative design solutions to the developer's problem. The essential point is not whether the architect is a co-owner, but whether he participates in the basic preliminary decisions that shape the project's design. As his chief professional benefit from sharing ownership, the architect can insist on his owner's right to help shape the project design.

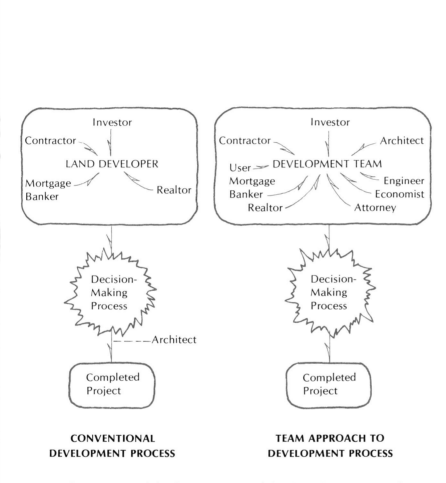

**CONVENTIONAL
DEVELOPMENT PROCESS**

**TEAM APPROACH TO
DEVELOPMENT PROCESS**

In the conventional development process (left), the architect is a virtual outsider, left out of the decision stage. As a member of the development team (right), he helps to shape decisions that, in turn, shape his design.

Architectural participation in the decision stage largely determines the project's design quality. As co-owner—or as a consultant given design control early in the development process—the architect has his greatest opportunity to serve the public. But, retained to apply architectural cosmetics to a preconceived basic design, the architect may serve as a reluctant collaborator in environmental destruction. The architect's participation in the decision stage is no panacea for poor environmental design. But at the least, it can serve to avert the grosser forms of urban despoilation—the tasteless commercial strips with hot-dog stands, chrome-trimmed diners, used-car lots, automobile graveyards, discount stores, beflagged gas stations, and garish motels—all united in scenic discord by billboards and signs screaming for attention. It can help to avert the monotonous single-family subdivisions built on treeless deserts scraped bare by bulldozers.

One motivating factor for the architect to get into development work is that it can help him recruit new clients. The case of John Portman & Associates, of Atlanta, shows how. In the dual role of architect and developer for Atlanta's famed Peachtree Center hotel-office-retail complex, Portman built a reputation for sound economic design coupled with imaginative architecture. Portman's boldly conceived Regency-Hyatt House, with its highrise atrium concept, provoked a skeptical reaction from conservative hotel people who confidently predicted that the project would never fly. After the Regency-Hyatt's spectacular financial success, the Portman firm was retained for several major hotels—notably, San Francisco's Embarcadero Center Hotel and the Regency-Hyatt House at Chicago's O'Hare International Airport. The design control assured by its status as architect-developer of Atlanta's Regency-Hyatt House gave the Portman firm an outstanding opportunity to display its hotel-designing skill plus an obvious opportunity for sound, lucrative investments in good development projects.

The point is not whether the architect is a co-owner, but whether he participates in the basic preliminary decisions that shape the project's design.

Along with these positive reasons for architects to enter development work, there are some negative, defensive reasons. The architect who offers the services in the 1970s that he offered in the 1950s may follow the trail of the dinosaur. This is not to say that all architects must become developers. Development work is not for everyone. Temperamental, psychological, financial, and even technical obstacles bar many architects' entry into development work. Many architects may remain happy, content, and perhaps prosperous in their traditional design role. But the profession's future may be jeopardized if too many architects stand pat. Expansion of the architect's role may be a condition of survival. A host of specialist competitors—package builders, project feasibility analysts, space programmers, space planners, cost consultants, building system consultants—has appeared on the construction scene. These aggressive newcomers are chipping steadily away at the architect's work—actual and potential. In view of the accelerating evolutionary changes underway in the building industry, the conservative architects who face the future with equanimity are ostriches. Development work is one of several ways for an architect to expand his services. And expanded services appear to be a condition of survival for the architectural profession as a major force in the building industry.

ARCHITECTS UNBOUND   Misunderstandings account for the failure of many architects to exploit the new opportunities. Some fear that they lack the necessary expertise. The flood of inquiries received at AIA headquarters indicates that many architects believe that equity participation in their projects is unethical. Still others harbor a vague distaste for the entrepreneurial role with its public image of the conniving wheeler-dealer con man.

These objections can be readily answered. By working on a development team, an architect can ally himself with the required experts—mortgage bankers, realtors, tax accountants, cost consultants, lawyers, and contractors. He doesn't need to be a renaissance man, expert in all fields. Moreover, he doesn't have to dominate the development process merely to participate in it. Ronald S. Senseman, FAIA, another pioneer architect-developer, has said:

"Participating in development work changes an architect's professional outlook. Instead of passively waiting for others to initiate projects, the architect aggressively creates projects himself. Before our firm got into development work, I was inspired with a kind of awe for the developer's financial acumen and his bold action in launching projects. But after our firm got into development planning, I found that the only thing some developers have to offer is the ability to get money. We found them highly dependent on our advice, and not only on design. The architect's ability to visualize a project gives him a tremendous advantage in the planning stages of the development process. The same talent that *creates* projects can also *initiate* projects."

In the normal course of practicing his profession, an architect—almost subconsciously—may accumulate a vast pool of knowledge applicable to the development process. His mind may be brimming with ideas for satisfying community needs and with valuable knowledge about local market conditions, trends in local land values, and the availability of potentially valuable land. He may be an expert in the technique of negotiating zoning changes. Among his professional and business friends, there may be potential equity investors—even mortgage lenders. This reservoir of expertise may lie impatiently awaiting the architect's recognition and eventual use in a real live development project.

The idea for a development project can be triggered adventitiously. Deeter Richey Sippel Associates' proposed Mon Plaza project was conceived in a traffic jam, when DRSA president Russell O. Deeter, FAIA, sat in his crawling automobile, looking down from the Fort Pitt Bridge onto a wasted, 48-acre riverside strip directly across the Monongahela River from

Pittsburgh's Golden Triangle. In place of warehouses, scrapyards, and parking lots occupying this superbly located site, Deeter visualized a thriving residential-commercial complex extending Pittsburgh's central business district. Later, questioned in a local television interview for ideas to improve the city, Deeter outlined his plan for rebuilding the wasted site.

DRSA initiated the proposed project by joining with a real estate company and a finance company incorporated as Mon Plaza, Inc., with each owning a percentage of the stock. The development team integrates the expertise in three key elements of project development—land, money, and design. The real estate member identifies potential markets, absorption rates, and possible major tenants. The financial member finds sources of front-end money, capital financing, and available government programs. The architect handles the physical planning, design, and construction management.

Current plans for the Mon Plaza complex include office buildings, apartment towers, a hotel, shopping center, marina, heliport, and rapid transit station, plus a 12,000-car parking garage, entered at an elevated street level and descending to the general site elevation. A proposed double-decked pedestrian bridge featuring a mechanical people-mover would link Mon Plaza with Gateway Center, about 900 feet away.

Inhibitions inspired by fear of violating the professional standards of ethics spring from misunderstandings. In an interpretation published in 1971, the AIA Board unequivocally laid this fear to rest: "As a participating owner of a project . . . [an architect] may perform in any role consistent with the position of ownership." As brief ethical guidelines, architects can protect their professional status—avoiding real and apparent conflicts of interest— by observing two basic rules: Inform clients and others who should know of any personal financial interest (apart from professional compensation) in the project. Contractually separate the design consultant function from the development team.

Architects as entrepreneurs are spanning the nation. Active in this work are RTKL, Inc., of Baltimore; Vincent G. Kling & Partners, of Philadelphia; Charles Luckman & Associates, of Los Angeles; and others. Smaller and newer architectural firms also are entering development work. One notably successful architect-developer is John T. Law, AIA, of Law Woodson Hammond, Palo Alto architects-planners, who started with a small restaurant-bar renovation. He took a piece of the action. Law since has gone into moderate-sized residential projects in the dual role of architect and co-owner.

The number of architects engaged in development work appears to be increasing rapidly. According to lawyer-developer Paul B. Farrell, Jr., a recent survey of nearly 100 firms responsible for some $4-billion annual construction volume indicates that one-third have been involved as principals in at least one development project. Another third indicated similar plans for the immediate future. In Farrell's words:

". . . As principals these architects were not merely consultants, but also took a 'piece of the action' [an equity ownership position] either in lieu of or in addition to their fee. Moreover, it is . . . clear that these architects

A traffic jam which trapped Pittsburgh Architect Russell O. Deeter, FAIA, on a bridge over-looking a wasted 48-acre riverside strip directly across from Pittsburgh's Golden Triangle fostered the proposed Mon Plaza Center. Current plans call for office buildings, apartment towers, a marina, a heliport and shopping facilities. A proposed double deck pedestrian bridge with a mechanical people mover would link Mon Plaza with Gateway Center across the river.

were not passive investor-consultants, but rather active decision-makers and co-developers."

To some extent, the architect's reluctance to assume the owner's responsibilities—and prerogatives—stems from the Beaux Arts tradition of architectural education, a psychic albatross dragged around by conservative architects. Whatever justification this concept had in the 19th century has disappeared in the latter third of the 20th. In this technologically dominated, fast-changing era, cost control, construction speed, market analysis, and financing have become ever more vital factors in the development process. In addition to design, modern architectural practice must embrace a host of other skills. The designer's talent still must be cultivated by the architectural profession, but not to the exclusion of the proliferating ancillary skills required to cope with modern building construction and development techniques.

While architectural education often has lagged behind building industry changes, many active practitioners have expanded their traditional roles as designers into larger roles as construction industry coordinators. This new role answers a need stemming from:

■ Replacement of individual owner clients by more sophisticated institutional clients demanding ever expanded services through blurred, bureaucratic lines of communication that have destroyed the old artist-patron analog of the architect-client relationship.

■ Growing complexity of building design, formerly handled by the architect with one or two engineering consultants, but now requiring a team of specialists to cope with new technological and managerial complexities.

■ Skyrocketing construction costs, which have made project economic feasibility highly dependent on cost control and fast construction.

BECKONING OPPORTUNITIES   Opportunities for architects to get a piece of the action never were greater. Inflation has evaporated a large part of the perennially shallow pool of interest-bearing mortgage funds available for development financing. Developers have sought new sources of financing, through new financing techniques designed to lure equity capital into development projects. Through professional involvement in the development process, architects have an exceptional opportunity to become joint venturers with developers, contractors, mortgage bankers, lawyers, realtors, accountants, and other experts contributing to the development process. Like corporate stockholders, they may take all or part of their professional compensation in equity shares in the project. But, to repeat, the architect's sharing of the entrepreneurial role is not essential to his participation in the crucial, early decisions of the development process. It does, however, assure his participation in those decisions.

Another factor encouraging architects' participation in development work is the shared experience of architectural firms that have opened this new frontier during the last 10 or 15 years. It no longer is pioneer territory. Today's architects can profit from the mistakes of the pioneers. There is ample room for individual experimentation in assembling development teams and assigning roles. John Portman & Associates, for example, never brings in a general contractor as a partner in a development project, preferring to preserve an arm's-length relationship between architect and contractor. Other successful architect-developers find contractors excellent partners. There also are many formulas for apportioning an architect's equity share, and an apparently limitless combination of roles he can play, provided only that he is qualified. Some architects function as developer, market analyst, financial analyst, architect-engineer, construction manager and—after project completion—building manager. But despite the tremendous area suitable for individual experimentation, some lessons learned at great anguish and financial expense suggest several general guidelines for neophytes:

Start cautiously. Wade, do not dive, into development work, constantly testing the water temperature. Along with opportunities, development work offers financial risks. The architect who boldly plunges into the biggest game in town for his initial development project may drown—financially. By limiting his equity participation to 20 percent of his professional compensation, an architect at best can ensure that he will break even, or at worst suffer relatively slight losses on a development project. Even 10 percent to 15 percent of the equity investment greatly enhances the architect's chances of controlling the design.

Do not compromise on professional compensation. Equity participation in a development project puts the architect's "blood on the line," as Vincent G. Kling, FAIA, put it. As members of development teams, architects, engineers, and planners run the risk of having their contributions undervalued in comparison to the contractors, developers, realtors, lawyers, accountants, and other business-oriented team members. If the developer attempts to justify a low architectural fee as the price of an alluring equity-sharing opportunity, withdraw from the project. It may be the *biggest* game in town, but it can't be the *only* game in town.

Ally yourself with competent, trustworthy partners. Set high standards for integrity. Consider compatibility with prospective joint venturers. A development project's owners are committed to a long-term association that is inevitably more intimate than the more formal architect-client relationship. It is one thing to have a so-and-so for a client and quite another to be married to him as a business partner.

Attain at least some elementary comprehension of market and financial analysis and financing techniques—enough to make a basic independent review of the project's financial feasibility.

Investigate all aspects of equity participation—professional liability, possible ethical conflict arising from a peculiar aspect of the deal, financial investment prospects—*before* involving yourself.

The architectural profession faces an uncertain future, ominous with threats from non-professional competitors, yet brimming with new opportunities. The entrepreneurial role in project development is one of the most promising of these opportunities.

THE TEAM APPROACH
TO THE DEVELOPMENT PROCESS

"The prima donnas have had their day. No matter how big the genius prima donna is, he cannot cover the areas necessary to solve the complex problems of urbanization. It takes a team."   —William W. Caudill, FAIA

Despite variations in the detailed arrangements for executing development projects, the process itself comprises three distinct stages—decision, design and delivery. In the decision stage, the owner decides what, where and when to build; in the design stage, the architect designs the project; and in the delivery stage, the contractor builds it.

The development process may be broken down this way:

Decision Stage
Assembly of development team, including owners
Economic design (market and feasibility studies)
Site selection
Program and budget
Construction cost projections
Schematic design
Land acquisition (option, lease or purchase)
Preparation of financing package for submission to
   prospective lenders

Design Stage
Preliminary design

          Financing negotiations (for mortgage and construction
             loans)
          Final design, construction documents
      Delivery Stage
          Construction
          Investment management—property management/leasing;
             sales or exchanges, liquidation of investment

One way or another, the foregoing tasks are performed in any competent approach to the development process. Remember, however, that "it ain't *what* you do, it's the *way* you do it." In today's high-powered construction world, the traditional building process is both too slow and too uncertain in its costs for sound investment planning. This chapter will explain why drastic changes are underway in the building industry. It will illustrate these changes in general, leaving detailed elaboration for later chapters.

Viewed from the broadest perspective, the team approach to the development process constitutes a modernization of the traditional building process. In every respect—technological, administrative, financial, economics and legal—today's construction world is more complex than yesterday's. The team approach exploits modern management techniques—originated for more technologically sophisticated industries—such as Critical Path Method (CPM), the Precedence Diagramming Method (PDM), and the Program Evaluation and Review Technique (PERT). Experts in these management techniques and in other fields launch an integrated attack on the formidable problems confronting today's developers. Because mortgage rates fluctuate, today's developers must continue to be more adroit than their predecessors in coaxing profits from their projects. Skyrocketing construction costs, combined with high interest rates, have made construction speed not merely beneficial but indispensable. The hard-worked principle of financial leverage (i.e., minimizing the owner's cash investment to maximize profit-equity ratios) adds risk as well as profit potential to the development game. Today success depends more than ever on accurate market studies, cost projections, favorable financing terms, fast construction schedules, and good coordination.

FLAWS IN THE TRADITIONAL BUILDING PROCESS   Historically, the development process has been a linear sequence of operations. Owner, architect, contractor and building manufacturers all played narrowly defined roles; decision, design and delivery stages were segregated and scheduled in neat chronological order with little or no overlap between the stages and little consultation among team members. Such a process was suited to the slower-paced handcraft era of the construction industry. But this fragmented process is totally inadequate for the evolving industrialized building process. The importance of good design planning, production scheduling, construction scheduling, purchasing, and coordination of scores of specialty construction trades requires a more tightly organized, more rigidly controlled and more intelligently planned process.

AVERAGE BUILDING COST INDEX

Index by Years

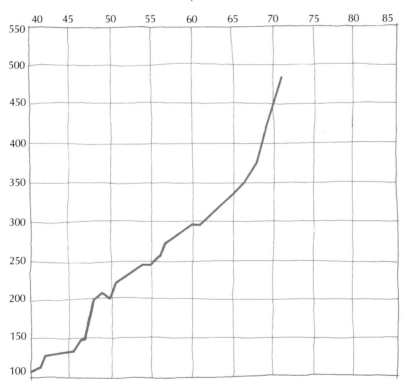

Haines Lundberg & Waehler/Architects
2 Park Avenue New York

The chief owner complaints against the traditional building process are lack of cost control and delays in completing construction. The traditional architect-contractor approach often lacks a formal mechanism for monitoring and controlling costs in the early project stages. Despite the architect's best efforts to obtain realistic cost estimates, the owner may suffer traumatic shock when the bids come in. Low bids on some recent projects have exceeded the budget by 25 to 50 percent. Moreover, in the old lump-sum bidding procedures used mainly by public agencies, costly delays are built into the process: several months may elapse between the completion of final design and the start of construction.

Success today depends far more on accurate market studies, cost projections, favorable financing terms, fast construction schedules, and good management than it ever has in the past.

Cost control and delays are not the only problems with the traditional building process. Some owners complain of poor construction quality and the lack of responsibility assumed by building products manufacturers. Others object to cumbersome administrative procedures and interminable disputes arising from contractors' claims for extras. As a consequence, the traditional building process, with its many communications gaps, its fragmented responsibilities, and its slow pace of execution slowly is yielding to the team approach, promoted by architects and other design professionals seeking better ways to serve their clients.

The traditional process fails to satisfy because it doesn't recognize the interdependence of the many contributors to a successful project and the need to rely upon their expertise during the decision stage. This urgent need for coordination springs from four specific factors:

- Industrialization of building technology
- Increased size of development projects
- Increasing need for good economic planning to assure financial success
- Growing economic importance of construction speed

Industrialization of building components demands far more planning than the old handicraft building process. When buildings were almost totally site-fabricated from wood, masonry, structural steel and reinforced concrete—and equipped with only primitive mechanical and electrical subsystems—preliminary planning was more simple. But as more sophisticated factory-fabricated building products enter the market, the lead time for advance production scheduling and preliminary architectural planning expands tremendously. Mechanical and electrical equipment, which may account for half the construction budget, have set the pace for industrialization. Even the building shell is becoming industrialized. Concrete masonry units for partitions can be ordered for delivery in a matter of days or weeks. But factory-fabricated demountable steel office partitions, designed to fit a modular floor plan, may require several months' notice to fit into a manufacturer's production schedule. Structural, mechanical and acoustical engineers and other technical specialists must be consulted during the decision stage. Otherwise, costly preliminary decisions may become locked into the project design, and opportunities for cost control may be overlooked.

The increased size of development projects vastly complicates real estate planning, increasing the need for more experts' participation in the early part of the decision stage. Shopping centers illustrate the trend. During the 1950s, shopping centers generally were restricted to small retail hubs, with one or two department stores, and a few shops and restaurants. Today, they are miniature central business districts with several department stores, scores of smaller stores and services, hotels, office buildings, theaters, apartments—even churches.

Successful planning of these giant projects is sophisticated. Land acquisition, local zoning problems, and planning for future traffic impact become increasingly complex. New towns and planned-unit development projects may require zoning variances that take years to extract from local zoning boards whose decisions often are colored by the clamor of suspicious, conservative citizens. Lawyers, planners, architects, engineers, accountants—and sociologists and pollution-control experts—may play vital roles in launching a project.

Good economic planning of development projects has become increasingly important in recent years. Financial planning of development projects has also become enormously complicated. The skills of the mortgage banker in arranging financing have become paramount. Financing often is the key to successful project development, and a host of techniques designed for the mutual benefit of borrower and lender has been devised to loosen the tight mortgage money market.

Inflation in construction and financing costs has placed a premium on speed in the development process. Once the developer has decided to build, each month's delay is a month's lost rent. But this is only part of the story; the financial penalties for delay are much greater than mere loss of income. Most important is the effect of delays on the developer's equity cash requirements. By cutting design-construction time from 24 to 18 months, and thereby generating early cash-flow receipts from a building, a developer may cut his equity cash participation by as much as 30 percent. This cash-investment reduction increases financial leverage, or profit-equity ratio, and leverage is the key to financially successful development (see time-saving graph next page).

At worst, lagging schedules can jeopardize a project's entire financial basis. Construction and mortgage loans generally are committed for definite time intervals. If prescribed conditions are not met before deadlines, the commitments become void. Mortgage loans often are limited by "rental attainment provisions" which set definite occupancy requirements as a condition for receiving mortgage loan balances. Design or construction delays can bring rental delays that threaten the developer's success in achieving the required occupancy level. Loss of 20 percent or even 10 percent of his permanent mortgage commitment—say, $100,000 on a $1-million loan—can turn an otherwise profitable venture into financial disaster.

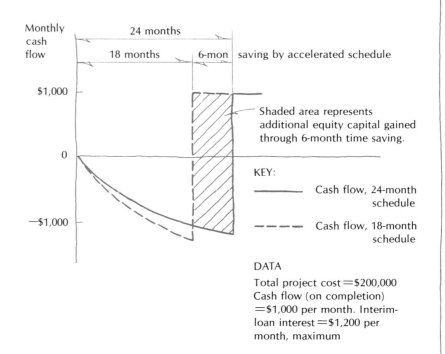

Monthly cash flow

24 months

18 months | 6-mon | saving by accelerated schedule

$1,000

Shaded area represents additional equity capital gained through 6-month time saving.

0

KEY:

——————— Cash flow, 24-month schedule

– – – – – Cash flow, 18-month schedule

−$1,000

DATA

Total project cost =$200,000
Cash flow (on completion) =$1,000 per month. Interim-loan interest =$1,200 per month, maximum

TIME SAVINGS MULTIPLY FINANCIAL LEVERAGE

An early occupancy date can substantially increase a developer's financial leverage. On the above project, interim loan interest gradually rises to $1,200 per month on a 24-month development schedule, as the contractor's progress payments fall due. This interest charge will rise at a slightly higher rate if the construction schedule is accelerated (see dashed line in above graph). But a 6-month earlier occupancy date can save $7,000 in interest charges and add $6,000 in cash flow, thus reducing the developer's equity requirement by $13,000. If the total equity for a 24-month schedule were $50,000, the additional $13,000 in effective cash flow would reduce the developer's equity by roughly 25%, and increase his financial leverage by about 30%.

Construction delays were a major factor in the 1968 decline and fall of the colossal Webb & Knapp real estate empire. The company's famous chief, William Zeckendorf, explains the collapse:

". . . a great many of our urban-redevelopment projects were running late and costing us money. . . . Less obvious, but in the end more painful, these projects had fallen out of phase with local markets because of their delayed construction cycles. Though in case after case we had been among the first to see the demand for quality housing in city cores, the delays we ran into in our pioneer efforts, and our own financial difficulties, permitted a wave of follow-the-leader builders to finish rival structures about the time we came to market. This competition meant great difficulties in getting adequate first rentals for many of our projects. As a result, we had losses, rather than income, on our books during those stretched-out beginning periods."

Delays are costly even on projects whose benefits are not measured in strictly economic terms. In past years, it sometimes was economical to delay the installation of site utilities to save the interest charges on the earlier payment of the capital cost. But with local construction costs escalating at 12 percent a year in many areas during the late 1960s and early 1970s, this strategy has backfired. For the past four or five years, it almost always has been more economical to build earlier rather than later. Optimists forecasting a leveling of future construction costs perennially have been disappointed.

THE GROWING COMPLEXITY OF THE DECISION STAGE In recent years, with the advent of the team approach, the decision stage has grown tremendously in relative importance. The growing complexity of the development process has made the design and delivery stages ever more dependent on controls developed during the decision stage.

Establishing these decision-stage controls is the developer's responsibility. The developer is the project's prime mover—a combination coach-quarterback of the development process. Behind the popular stereotype of the developer as a speculating wheeler-dealing promoter is his essential managerial function as the leader of the development team. The developer usually is at least part owner, but ownership is not an intrinsic part of the developer's role. He may be a real estate broker, a mortgage banker, a contractor, or an architect retained by an owner whose own part in the development process is a purely passive investor's role. Coordinating the work of the decision stage is the developer's chief responsibility. After he pushes the project through the decision stage, he can rely on momentum to carry it through the design and delivery stages. If he has done his work well during the decision stage, the developer can become more a monitor than a driver during design and delivery.

Once he has made even a tentative decision to go ahead, the developer should start assembling the development team at the earliest practicable date. He will find immediate need for most of the team members' talents.

In addition to the developer, a team should contain:

■ Design professional—architect, engineer, or planner and his own team of consulting design specialists: landscape architect, acoustical consultant, other technical experts.

■ Attorney, responsible for reviewing legal documents for the team's legal structure, project financing, consultants' services, land purchase, leases, and zoning changes.

■ Mortgage banker, responsible for arranging financing (mortgage and construction loans).

This is only one of thousands of possible combinations. The development teams may be expanded to include a tax accountant, an insurance broker, a market analyst, a contractor and a real estate broker.

The key development tasks during the decision stage are:

■ Market, feasibility studies, ■ Land acquisition ■ Project financing.

The sequence of other decision-stage tasks may vary. Assembly of the development team, for example, can occur either before or after preparation of feasibility studies. But unless the developer already owns or controls the site, the major steps should follow in the indicated order: market and feasibility studies should precede land acquisition, and land acquisition should precede submission of the financing package to prospective lenders.

The major part of the economic design of a development project is performed during the decision stage. A *market study* initiates the process. The market analyst first assesses the prospects for a project in a given area. If the analyst concludes that there is, in fact, an economic need for a new hotel, office building, nursing home, shopping center, or industrial park, he concludes with recommendations to build x area of leasable space at a rental of y dollars per square feet. The later *feasibility study* establishes the maximum total project cost, or budget, that can yield the desired rate of return on the developer's cash investment. The developer's decision to build depends on a construction cost estimate limited within the project budget. Economic design and physical design thus proceed more or less concurrently with continual interplay between the two parallel processes.

Once he has made even a tentative decision to go ahead, the developer should start assembling the development team at the earliest practicable date.

Land acquisition is the next major decision-stage step—a step that normally is beyond the architect's experience. It proceeds more or less concurrently with several intervening steps—programming, budgeting, construction cost projection, and schematic design. Land acquisition is the tough follow-through step to site selection which is indicated, if not dictated, by the market study. Tying up the land, through lease, purchase or option, is a prerequisite for arranging the project's financing.

As the final step in the decision stage, the developer must arrange for *project financing*. On a conventionally financed project, the two basic loans are a mortgage (permanent) loan, advanced after the project is completed, and a construction (interim) loan, advanced at the start of construction. The key loan is the mortgage loan. Without it, or some alternative form of permanent financing, no prudent lender will advance a construction loan, for the good reason that his loan is paid off by the mortgage lender. Thus the developer's primary efforts are directed toward permanent financing—i.e., a mortgage loan, sale-leaseback, or some other form of financing designed to keep the project financially afloat through a major part of its assumed economic life. Once the permanent financing is arranged, the construction loan can be readily negotiated.

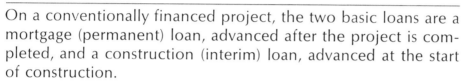

On a conventionally financed project, the two basic loans are a mortgage (permanent) loan, advanced after the project is completed, and a construction (interim) loan, advanced at the start of construction.

As a first step in securing a mortgage loan, the developer prepares a mortgage (or financial) package for prospective lenders. It contains the basic data needed by the lender to evaluate the project as an investment vehicle. It includes summaries of the market and feasibility studies; the developer's economic projections, including equity investment and first-mortgage requirements; a complete functional and technical description, illustrated with rendering, site plan, and floor plans; and a description of the lessees. Several other documents—notably the owners' individual financial statements and lessees' leasing agreements—are combined with the mortgage package presented to prospective lenders. Compiling this package is a major duty of the mortgage banker, who also takes care of the formal application, deposits, final letter of submission, and other formal duties associated with negotiating the loan. Once he obtains a lender's mortgage commitment—i.e., a contract to advance a mortgage loan subject to successful completion of construction—the mortgage banker can negotiate a construction loan for the developer.

Preparation of the financing package is the final, crucial task of the decision stage, essential to pump financial lifeblood into the infant project. Without a good financing package, capable of convincing a prospective lender of its sound physical and economic design, a private development project probably will fail.

THE DESIGN STAGE   From the drastic changes effected in the decision stage, differences between the traditional building process and the modern team approach diminish in the design stage. The principles of good architectural design and the basic methods of implementing them obviously remain unchanged. From the developer's viewpoint, the key concern during the design stage is cost control. It depends upon cooperation between the developer and the architect—the key team members during the design stage. Good relations between these two are essential to assure that the project fulfills its function at an acceptable cost.

THE DELIVERY STAGE   Just as a good financing package is vital to a project's birth, good management is essential to its continued survival and health. The traditional building process, with its possibly antagonistic architect-contractor relations, often impedes efficient management and saps economic vitality in the delivery stage.

The package builder provided a non-professional answer to several ills in the old architect-general contractor approach. By offering a combination of design services *and* construction, the package builder solved the problems of early cost control and speed. In the package builder's integrated organization, the estimating department could monitor the cost of the designer's product continuously. The estimator could recommend alternatives early in the planning process, before the project got locked into final shape. He also could schedule the preliminary design process to start construction long before the final design was completed. Site grading and foundation construction could start several months before the selection of door hardware. And by unifying the design-build function, the package builder offered owners a simpler administrative mechanism for delivering buildings.

For most owners, however, the package-building approach contains a fatal flaw—loss of independent professional design services. For most commercial, institutional, industrial and governmental projects, the lack of an independent architect, and his consultants, constitutes an intolerable sacrifice of design quality. Simply stated the problem is: How do you combine the cost control and speed achieved by the package builder with the design quality achieved under the traditional architect-general contractor approach?

Several modifications of the traditional building process have provided interim solutions to these problems. Lump-sum, general-contract bidding on complete drawings and specifications is becoming obsolete in private industry. It is both too time-consuming and too economically uncertain for most industries to tolerate. Moreover, the increasing scale of building projects has reduced the number of general contractors financially qualified to bid. Various types of negotiated contracts—cost-plus a percentage, cost-plus-fixed fee, or lump-sum bids on different parts of the construction—allow early construction starts. The contractor sometimes is engaged as a consultant to the architect on costs and construction technology. Owner-architect-contractor teams often work together throughout the entire development process, like an ad-hoc package-building firm. Such an arrangement constitutes a tremendous improvement over the traditional building process, with the architect and later the contractor entering the scene at the last possible moment, too late for full use of their talents. It also removes the biggest problem with the package-builder approach—the loss of professionalism in the design function. But even the owner-architect-contractor team, though admirably suited to many kinds of development projects, still constitutes an expedient rather than a general industry solution. For large projects—i.e., projects whose total construction cost exceeds $5 million to $10 million—the most promising solution advanced to date is the construction manager.

THE CONSTRUCTION MANAGER   The advent of the construction manager is one of the biggest evolutionary changes now underway in the building industry. The construction-management approach constitutes a direct attack on the two major flaws in the conventional construction process—lack of cost control and excessive time consumed by the serial sequence of decision, design and delivery.

Viewed in a historical context, the construction manager represents a professionalization and expansion of the general contractor's former entrepreneurial-managerial role. Unlike the traditional general contractor, the construction manager participates in the decision and design stages as well as the delivery stage. According to Philip J. Meathe, FAIA, president of Smith, Hinchman & Grylls Associates, Inc., among those potentially qualified to assume the construction manager's role are architects, engineers, general contractors, planners, real estate developers and management experts.

The construction manager approach responds to growing public and private dissatisfaction with the general contracting approach. There has been dissatisfaction with the practice of bid shopping—i.e., submitting bids based on the bidder's own estimates and then, if he wins the award, tending to squeeze subcontractors' prices to maximize his own profits. Not surprisingly, this practice never has enhanced construction quality. Disenchanted owners, seeking remedies to this and other previously discussed deficiencies in the traditional building process, have abandoned the general contracting system for the new construction manager approach. The General Services Administration also is experimenting with this new approach; it appears to be the wave of the future.

The construction manager's key contribution is rigid control over costs and time. In a sense, he bridges the gap between architect and contractor.

The construction manager's key contribution is rigid control of costs and time. In a sense, he supplies the formerly missing link between architect and contractor.

His skills are required throughout the entire development process—for cost estimating during the decision and design stages and for scheduling during the delivery stage. His job combines the general contractor's basic management role with some duties traditionally handled by the architect. Like other development team professionals, the construction manager acts as the owner's agent. Unlike the general contractor, however, he gains no profit from cutting corners on quality.

The professional construction manager should not be confused with the project administrator. As a member of the owner's organization, the project administrator (or project manager) is the client's voice and purse string, signaling the project start and acting as the owner's liaison with the other members of the development team.

The construction manager serves in a line rather than a staff capacity. His duties include: ■ assembly of construction documents for bids, ■ solicitation of bids from contractors, ■ negotiation with successful bidders, ■ preparation of construction schedules, ■ negotiation of contractors' claims, ■ construction cost control and ■ processing of subcontractor requisitions and other papers required for payments, plus other supervisory and administrative work usually performed by the general contractor.

Who assumes the role of construction manager depends on experience, skill and, of course, enterprise. Architect-engineer firms that have acquired experienced construction personnel are qualified for this work. And many general contractors have abandoned the attitudes of the old carpenter-foreman-turned-contractor and modernized their traditional contractor skills to include computer-calculated Critical Path Methods, accelerated scheduling and other new construction management techniques. The construction manager's domain ranges from this sophisticated paper work to head knocking at job-site meetings. If, for example, the concrete contractor does not allow the plumbers and electricians adequate time to place their conduit, inserts, and sleeves before a concrete pour is scheduled, the construction manager must resolve the dispute.

ACCELERATED SCHEDULING   Basically, however, the construction manager's job is to apply scientific management to the development process so that crucial decisions are made at the required time to keep the project rolling on schedule. Because it takes much longer to build a building than to design it, the key to accelerating the overall construction process is to telescope design and construction—to start foundation construction as soon as possible after completion of a minimum amount of basic design (see chart). Information required to build foundations includes column spacings and loads and foundation wall locations. Thus the structural framing scheme must be determined early in the design stage. This basic structural design becomes a hypercritical item on the "critical path"—i.e., the sequence of activities that controls the time for completion. But it is not necessary to have every last column-truss connection detailed. What the structural engineer, architect and construction manager must know at this preliminary stage is the certain availability, at an acceptable price, of a structural framing system that will satisfy the preliminary design scheme.

The price of the accelerated schedule, with its overlapping, coordinated project stages, is early commitment. With accelerated scheduling, you lose the luxury of second-guessing your basic design decisions. You must commit

Decision	Design	Delivery
Program & Budget	Design & Construction Documents	Construction

CONVENTIONAL LINEAR DEVELOPMENT PROCESS

Telescoped Schedule   Time saving

Staged contracts

ACCELERATED SCHEDULING

Accelerated scheduling telescopes the development process into a faster, more tightly organized operation, requiring earlier commitment and tighter coordination as the price for potentially great time and cost savings.

yourself to build long before the final contract is let. Programming and design must proceed in a more rigorously controlled way. In the traditional building process, communications gaps among owner, architect, manufacturers and contractor may be merely costly. In an accelerated project, they can be disastrous.

Acceleration of the development process achieved through telescoping of the decision, design and delivery stages is one of the basic features distinguishing the modern team approach from the traditional development process. Another is the trend toward greater participation in the decision stage by all development team members whose impact may influence early decisions. This philosophy was illustrated by Robert Allan Class in the AIA Journal in the following hypothetical case study.

CASE STUDY—CANCER RESEARCH LABORATORY   The idea for the project originates at a Chamber of Commerce luncheon in Eupolis, Neb. (100,000 population), where a federal grant for cancer research is announced. The recipient is Purepills, Inc., a large AA drug company lured to the city five years earlier by the local chamber and the city's development coordinator. Meeting at the luncheon, an architect, a banker, and a realtor—discussing news of Purepills' imminent need for laboratory space—come up with an idea for a joint venture.

The company will require new laboratory space within 18 months. In the meantime, it can lease more limited quantities of additional space outside its existing facilities. The program features expansion of a current 500-man R & D staff to 1000 at the end of the first year, to 1500 by the end of the second year, and to 2500 by the end of the third year. At this level, the staff is expected to remain static for 10 years, but the program must allow for a possible further expansion, depending on the results of the research.

As it currently is constituted, the building industry in America cannot handle the vast volume of construction needed by the year 2000.

Funding for the expansion is reasonably secure. Congressional pressure has prompted the President and the Secretary of Health, Education and Welfare to announce a massive 10-year program for lung-cancer research, and Congress has passed a resolution supporting the program, giving good assurance for continued federal funding. HEW has decided on private research, with annually established goals and programmed personnel needs. Purepills has won the bulk of the HEW research grant after submitting a proposal in competition with other private companies. In addition to its prospects for continued federal financing, Purepills has indications of substantial additional money from private foundations.

Inspired by the local chamber and the city development coordinator's

PROJECT FINANCE ANALYSIS
(From Computer Printout)

BUILDING   A CANCER RESEARCH LABORATORY
PROJECT NO.   9105
LOCATION   A MIDWEST CITY OF 100,000 POPULATION, EUPOLIS, NEBR.
OWNER   JOINT VENTURE
BASIS OF EST. 2,500 EMPLOYEES 2,000 PARKING SPACES   49 ACRES @ $2,095
BASIS OF EST. GROSS RENTABLE BASIS NET-NET LEASE
DATE   3-4-70

PROJECT COST VALUE	S.F. GROSS AREA	PERCENT	S.F. DOLLAR	TOTAL DOLLARS	UNIT COST
001 GROSS BUILDING	641,667	100.00	$ 24.00	$ 15,400,008	
002 TENANT ALLOWANCE		0.00	0.00	0	
003 LAND		0.66	0.16	102,666	
AND IMPROVEMENTS 1		3.45	0.83	532,583	
004 FIELD COSTS 2		104.11	$ 24.99	$ 16,035,258	
005 DEVELOPER EXPENSE 3		24.12	5.79	3,715,251	
006 NO MISCELLANEOUS COSTS		0.00	0.00	0	
007 PROJECT COST VALUE (S.F. GROSS)	641,667	128.23	$ 30.78	$ 19,750,510	
(UNITS)	1				19,750,510

PROJECT INCOME VALUE					
008 NET RENTABLE AREA - GROSS INCOME	641,667	100.00	$ 3.71	$ 2,380,584	
009 AVERAGE OCCUPANCY AREA - GI	641,667	100.00	$ 3.71	$ 2,380,584	
010 TOTAL BUILDING AREA - GI	641,667	100.00	$ 3.71	$ 2,380,584	
011 TAXES 4		0.00	0.00	0	
012 OPERATIONS 5		0.00	0.00	0	
013 NET INCOME FOR DEBT-DEP-EQUITY		100.00	$ 3.71	$ 2,380,584	
014 LESS % RETURN ON LAND		8.00		8,213	
016 NET INCOME BLDG BEFORE DEPRECIATION				$ 2,372,370	
017 CAPITALIZED AT ---------- .1000 (12.90% - 15.00 YR.)				$ 18,390,465	
018 LAND AT COST				102,666	
019 PROJECT INCOME VALUE (S.F. NET)	641,667	100.00	28.82	18,493,131	
(UNITS)	1				18,493,131

MORTGAGE EQUITY				
020 MORTGAGE ON PROJECT COST VALUE	80.00		15,800,408
021 OWNERS EQUITY ON PCV	20.00		3,950,102
	100.00		$ 19,750,510
022 ANNUAL AMORTIZATION ----- .1000 (12.90% - 15.00 YR.)			2,038,252
023 FIRST YEAR INTEREST			1,580,040
024 FIRST YEAR PRINCIPAL PAYMENT			458,211
025 AMORTIZATION FOR LOAN YEARS			30,573,789
026 INTEREST ON LOAN FOR LOAN YEARS			14,773,381
027 PRINCIPAL ON LOAN			15,800,408
028 NET TAXES AND OPERATIONS	$ 0		
029 AMORTIZATION	2,038,252		
030 ANNUAL RETURN			
ON INCOME	14.38		342,331
ON EQUITY	8.66		342,331

TAXABLE INCOME	ONE YEAR	15.00 YEARS
032 INCOME	2,380,584	35,708,760
033 INTEREST	1,580,040	14,773,381
034 TAXES	0	0
035 OPERATIONS	0	0
036 DEPRECIATION ON STRAIGHT LINE METHOD	589,435	8,841,529
PROJECT COST VALUE LESS LAND $ 19,647,844 @ 3%.		
037 EXPENSE	2,169,475	23,614,910
038 TAXABLE INCOME	211,108	12,093,849
039 AVERAGE ANNUAL TAXABLE INCOME		$ 806,256

DEPRECIATION RECOVERY	PERCENT	12.00 YEARS	25.00 YEARS
040 STRAIGHT LINE METHOD	36.36	$ 7,143,956	$ 19,647,844
041 200% DECLINING BALANCE	52.77	10,368,167	19,647,844
042 SUM OF DIGITS	58.82	11,556,861	19,647,844
043 LAND RESIDUAL		102,666	102,666
044 LAND APPRECIATION	10.00	10,267	102,666
045 INCOME		28,567,008	59,514,600
046 TOTAL RECOVERY		35,823,897	79,367,776
047 ANNUAL AVERAGE RECOVERY STR. LINE		$ 2,985,324	$ 3,174,711
048 ANNUAL AVERAGE RECOVERY 200% DEC		3,254,009	
049 ANNUAL AVERAGE RECOVERY SUM DIG		3,353,066	

1) Incl. Broker's Commission
2) Total of 001, 002, 003
3) Incl. Design Fees, Legal/Accounting Fees, Construction Loan Fees, (Incl. Broker's Commission and Finder's Fee), Economic Research Fee, Interest/Taxes/Insurance During Construction, Completion Bond, Surveys and Tests, Contingency, Developer's Profit and Broker's Fee for Permanent Financing.
4) Assumed by User
5) Assumed by User

From *AIA Journal* 7/70

efforts to assure construction of the new Purepills laboratory in Eupolis, the architect, banker and realtor decide to embark on a joint venture. Though they have had only minor business contacts with one another, they have become acquainted through service on the United Fund Appeal campaign and other civic works, and consequently share a mutual respect.

After their luncheon conversation, the joint venturers agree to make a proposal to build and lease an R & D laboratory for a 10-year period. Each partner will put up $10,000 to cover initial expenses, with each taking a one-third equity share. Each will contribute his professional expertise. Since each partner is a majority stockholder in his firm, he can call on staff assistance to the joint venture, to be reimbursed at cost.

Architects are applying their creative imaginations to the entire development process—conception, initiation, economic analysis, administration, financing, and even the post-construction management of development projects.

After a preliminary meeting with Purepills officials, the new team acts quickly to prepare a proposal to the drug company. The architect prepares a Critical Path Network of tasks needed to produce the proposal in 30 days. The realtor locates a land tract apparently suitable for the laboratory and open for an option. The architect does a preliminary site-use study, reviews zoning problems and prepares rough construction cost projections (see financial analysis chart). The banker makes economic analyses for the go-no-go decision. The team then recruits an attorney to review zoning and determine the best legal structure for the joint venture. As their final preliminary step, they submit an illustrated proposal to design, build and lease a cancer research laboratory to Purepills.

This fast action wins the project, largely because the accelerated team approach to project initiation has overwhelmed the competition from independent architects, who simply call to inquire if Purepills needs an architect. After signing a contract with Purepills, the joint venturers take out a $10,000 option to buy the site and start refining their rough preliminary studies for a crash development schedule that will produce the laboratory in the required time:

For the architect co-owner, this project offers:

- Guaranteed participation in the decision stage, including site selection
- Assurance of design control
- A good long-term investment in addition to professional compensation.

The architect prepared a simplified Critical Path Network analysis (see next page) of the necessary steps leading to the proposals.

CRITICAL PATH NETWORK FOR PREPARING PROPOSAL

Days

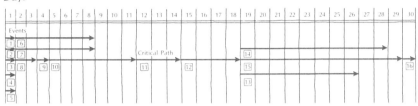

1. ESTABLISH CONCEPT AND
   BASIC FRAMEWORK OF
   JOINT VENTURE
   a. Determine Participants
   b. Evaluate Their Capabilities
      and Potential Input of
      Talent/Money/Land
   c. Select Attorney

2. HOLD PRELIMINARY MEET-
   ING WITH USER,
   DETERMINE:
   a. Agreeability to Consider
      Proposal from Joint Venture
   b. Concept of Project/
      Available Detail
   c. Type of Proposal: Full
      Lease/Full Purchase/
      Option/Other

3. DETERMINE APPROX.
   BUILDING SIZE AND COST

4. DETERMINE LAND
   REQUIREMENTS

5. PREPARE ROUGH ECO-
   NOMIC ANALYSIS FOR
   "GO-NO-GO" DECISION
   CONSIDER:
   a. All Economic/Time Factors
   b. Dollar Size of Project and
      Ability to Finance
   c. Tax Projections and Con-
      sequences to Investors
   d. Alternatives for Flexibility

6. SECURE LAND OPTIONS.
   CONSIDER:
   a. Zoning Implications

   b. Utilities
   c. Access
   d. Location
   e. Site Characteristics
   f. Tax Implications
   g. Price vs. Value
   h. Building Expansion
   i. Related Development

7. EXPLORE POSSIBLE MONEY
   SOURCES

8. PREPARE PRELIMINARY
   PROGRAM

9. RECHECK WITH USER

10. DEVELOP SCHEMATIC
    DESIGN FOR BUILDING
    AND SITE

11. PREPARE COST ESTIMATES
    FOR BUILDING AND SITE
    IMPROVEMENTS

12. REFINE ECONOMIC
    ANALYSIS AND DEVELOP
    NET RENT TO USER

13. CONFIRM FINANCING, AND
    a. Obtain Other Equity
       Partners as Needed
    b. Determine Cash/Land
       Input by Investors
    c. Reconfirm Tax
       Consequences

14. DECIDE ON LEGAL STRUC-
    TURE OF INVESTORS

15. PREPARE WRITTEN AND
    GRAPHIC PROPOSAL

16. PRESENT PROPOSAL TO
    USER

ECONOMIC DESIGN

"An easy way to compound the problems of the physical design is to let someone else make all the decisions relating to the economics of the project."
— Herbert E. Duncan, Jr., AIA

Economic design embraces the entire scope of project economics. It starts with the early market and feasibility studies, which form the analytical basis for the developer's decision to build and the lender's decision to finance the project. It concludes with an economic analysis projecting the owners' investment prospects some years into the future.

Economic design proceeds more or less concurrently with physical design, paralleling it at each stage of the development process (see economic and project design chart next page). Market and feasibility studies initiate the entire process. A tentative affirmation of feasibility is required to justify even preliminary design. When the feasibility study flashes a green light for the project, the physical design begins in earnest. While the architect is defining the project's general arrangement and space use, determining the building masses and performing other tasks associated with schematic design, the financial analyst—a tax attorney, accountant, mortgage banker, or possibly a member of the architect's own firm—should be studying investment strategy. Would the owners benefit most from a sale-leaseback deal

Market and Feasibility Studies	Budget	Analysis of developer's investment prospects, tax factors, cash flow	Financing Package

ECONOMIC DESIGN

Project Scope	Construction Cost Projection	Schematic Design	Constr. Cost Estimate	Construction Documents

PROJECT DESIGN

Economic Design and Project Design should pursue closely coordinated, parallel courses.

in which they would give up depreciation tax shelter in return for additional financing? Or should they go for the lesser financing of a conventional mortgage and retain the depreciation tax shelter? If so, when should they sell the project? Answering such questions is the financial analyst's job.

For speculative projects, the developer needs several studies:

■ A market study ■ Appraisal ■ Feasibility study ■ Investment analysis.

Like project design, economic design proceeds as a series of successive approximations—from the basic to the refined. Each of the basic studies— market study, appraisal, feasibility study—generates data needed for its successor. The market study is analogous to schematic design. And, as the final stage of the economic design, investment analysis is analogous to the final detailed design.

Economic design is far less sophisticated than project design, and accordingly is conducted on a less formal basis (see project and economic design charts). On large, complex projects, it may, however, become quite sophisticated. Added to the four basic studies, a *highest-and-best-use study* may follow the feasibility report. In such instances, economic design may generate a series of reports stacked a foot high. On a simple, small-scale project, there may be no formal studies at all; the needed data may fit on two pieces of scratch paper. Yet in one way or another, economic design should proceed through the same basic steps.

MARKET STUDY   The market study projects future supply and demand for a specific type of commercial, residential, or industrial project. Following this general study, the market study then focuses more narrowly on a specific property in a specific location. It should offer recommendations for financing, sales techniques, and amenities promoting successful leasing (or sale) of the space. It should recommend the volume of space to be rented or sold and the sale or rental price. On large, complex projects, this part of the market study may constitute a separate *marketability study*.

A market analyst assessing prospects for a hotel in Zenith will need answers to the following questions: How many hotels does Zenith currently have? How old are they? Are they profitable? Do they have convention facilities? Is Zenith potentially a good convention town? Will it support the relatively high room rates needed to justify construction of a new hotel?

When he selects a site, the market analyst will need to investigate transportation—current and planned future access to airports, general highway patterns and mass transit. Is the immediate district surrounding the proposed site decaying or improving? How will it fare in the future under the action of the dominant economic, social and political forces in the metropolitan area? Will growth in office buildings, government buildings, retail and entertainment facilities help to support the hotel? Or is this district destined for economic oblivion?

Market analysis for an apartment project can be more formalized, since it concerns a type of development in more general demand (see the apartment market analysis next page).

GUIDELINES FOR APARTMENT MARKET STUDY

I.   Research demographic data

1. Project population trends, city and metropolitan area
2. Review local housing situation—building type, unit sizes, rents, occupancy, age, and condition. (Obtain data from last two U.S. Census reports on housing —1960 and 1970)
3. Compare existing housing around proposed project site with general housing in city.

II. Survey recent apartment construction trend in city and metropolitan area
     Sources: Bureau of Census construction reports; local FHA office; local real estate board and home builders' associations

III. Survey city's existing rental market

1. Rent levels, by unit size and quality
2. Occupancy levels (also by unit size and quality)
3. Recent trends in apartment construction, by building type and apartment unit size

IV. Estimate overall rental apartment demand by (a) building type (highrise, garden, etc.) and (b) unit size (efficiency, 1, 2, 3-bedroom)

V. Recommend number of units and rentals, weighing these factors:

1. Prospective tenants' incomes
2. Apartment mix, based on estimated market demand
3. Amenities for each unit, by size and type
4. Comparable rents in similar housing throughout city (considering location, transportation, access to shopping, schools, etc.)

VI. Estimate "absorption" period for apartment units

APPRAISAL BY INCOME METHOD, ILLUSTRATIVE EXAMPLE

Assume the following data:

(1) Commercial office building with 25,000 sq ft *leasable* space.
(2) Rent = $8 per sq ft per annum
(3) Operating expenses (real estate taxes, insurance, utilities, management fees, maintenance, plus allowance for vacancies) =40% gross rental receipts
(4) Capitalization rate =10%

Step 1: Compute gross rental income =$8 x 25,000 =$200,000

Step 2: Operating expenses =40% x 200,000   = −80,000
                                 Net income =$120,000

Step 3: Compute capitalized value as follows:

$$\text{Capitalized value} = \frac{\text{Net income}}{\text{Capitalization rate}}$$

$$= \frac{\$120,000}{0.10}$$

$$= \$1.2 \text{ million}$$

Though the same firm or individual often performs all these economic studies, there are important differences between market analyses, on the one hand, and appraisals, feasibility studies, and investment analyses, on the other. On a small, single-purpose project, with relatively short-term goals, a local realtor who knows the local market for apartments, office, or industrial space may produce a good market analysis. But though perfectly competent to assess supply and demand, going rental rates, and sales prices, the realtor may be incapable of doing the more complex financial analysis. Market studies for complex, long-term, large-scale projects similarly may be beyond the local realtor's competence. In a new town, for example, heavy early investment in utilities and site work may produce cash-flow losses (i.e., negative net income) in the project's early years. A market study for such a long-term investment may require consultation with sociologists, city planners and ecologists.

Another warning on market analysis concerns timing. A rental project's investment prospects may depend on early success in leasing space. Failure to fulfill a minimum occupancy requirement, often needed to qualify a project for all or part of its committed mortgage loan, can present a developer with a sudden desperate need for a vast amount of equity capital merely to keep his project afloat. Through ad-hoc negotiation with the permanent or interim lender, the developer may save the project, possibly by surrendering an equity share to the lender. But no matter how well the project performs economically over the long run, a slow start can shatter the owners' investment prospects.

APPRAISAL   Before making his go-no-go decision, the developer needs an appraisal to assess his financial prospects. Normally, the developer first wants to know the maximum mortgage (permanent) loan for which his project can qualify. This mortgage loan is directly related to appraisal value. Most conventional mortgage loans are limited to 75 percent of total appraised value (land plus buildings and other capital improvements).

Architects and laymen sometimes are confused about the basis for mortgage appraisals, mistakenly assuming that they are based on project cost. Real estate investors are not interested in how much an income-producing property costs; they are interested in how much net income—cash flow— it will generate. Buildings have no intrinsic value to these investors. A marble-faced, palatial motel built in an economically dying rural area may be virtually worthless. A strategically located motel built at one-third the cost may be a potential uranium mine. Through some ingenious combination of systems building and accelerated scheduling, an office building might be built for a super-economical $20 per square foot. Or, through some grotesquely inefficient development process, the identical building might cost $50 per square foot. Either way, the appraised value is the same.

For an income-producing property, the most rational method of appraising a real estate project is the *income method*. The other two appraisal methods—*replacement cost* and *comparable sales*—may be helpful as a check. Comparable sales often can serve as the basis for setting sales prices on single-family residences. But for income-producing property, neither replacement cost nor comparable sales count for much; they are small factors in determining economic value.

To calculate an appraisal by the income method, the financial analyst needs the data produced by the market (or marketability) study—leasable area (in square feet or apartment rooms) and annual rental price (per square foot or apartment room). He also needs operating expenses (often estimated as a percentage of gross rent) and a *capitalization rate,* the percentage of net income (gross receipts minus total operating expenses) with respect to total project value. From these basic data, the financial analyst computes the capitalized value—or market value—which serves as a basis for calculating the mortgage loan (see appraisal by income chart).

In the early years of a project, depreciation may reduce income taxes and perhaps produce a paper loss shielding other income from taxation.

Appraised value, and consequently the mortgage loan amount, depends on the interest rate. The major factor in determining an acceptable capitalization rate is the interest rate on the mortgage loan—the higher the interest rate, the higher the capitalization rate. Since the capitalized value is *inversely* proportional to the capitalization rate, the effect of high interest rates is to *reduce* capitalized value and consequently the mortgage loan available. (A building appraised at $125,000 on the basis of an 8-percent capitalization rate drops to $100,000 appraisal at 10-percent capitalization rate.) A reduced mortgage loan, in turn, raises the cash investment required from the developer, thus reducing financial leverage or profit-equity ratio. High interest rates depress a development project's investment prospects ▪ by reducing the mortgage loan amount for which the project qualifies and ▪ by raising the owner's operating expenses.

FEASIBILITY STUDY   The most important economic study is the feasibility study. It normally forms the basis for the developer's decision on when and what to build. In its most common version, the feasibility study presents the developer with the key information he needs for making his decision— the maximum project cost, or budget, that can produce an acceptable return on the owner's cash investment.

For a feasibility study, the developer needs to know the project's appraised value; mortgage terms (loan amount, interest rate, term of constant annual payment); and minimum acceptable rate of return on equity investment. From these data, the financial analyst calculates the maximum project cost, or budget, that can yield the required return. If the cost control expert decides that the project can be built within the budget, then the project is feasible. (see feasibility study chart on page 36.)

In computing the project cost, the financial analyst must include all project costs: land cost, architect-engineer fees, construction inspection fees, interest on construction loan, real estate taxes during construction, filing fees, insurance, new equipment and furniture costs, moving expenses, and miscellaneous items not included in the construction contracts.

Keeping construction costs safely within the budgetary limit requires continual monitoring, starting with schematic design. Later, in the design stage, if costs have not been controlled during the decision stage, the architect may be forced to substitute painted concrete block for plastered partitions, aluminum entrance hardware and trim for stainless steel, resilient tile for terrazzo in the main entrance. Too many substitutions of cheaper building components could, of course, jeopardize the projected rental price. Throughout the decision and design stages, there must be continual interplay between economic design and project design. To assure this the architect must participate in the decision stage. Otherwise, costly time-consuming redesign may be required.

HIGHEST-AND-BEST-USE STUDY Far more complex and far less common than the feasibility study is the highest-and-best-use study. On highly valuable sites where several different types of projects might all prove economically feasible, one project would probably produce a maximum return on equity investment, if the developer were clever enough to identify it. This is what the highest-and-best-use study attempts to do.

In conducting a highest-and-best-use study, the financial analyst first uses a process of elimination, rejecting several potential land uses on the basis of his judgment. Then, having narrowed the field to the most likely prospective uses, he screens them for marketability and selects several for detailed feasibility studies. As a last step in this process, he may compare a 12- with a 20-story apartment tower, or even attempt a height-optimizing study that would identify precisely the most profitable story height for a given use. In so doing, the market analyst would work closely with the architect who, in turn, would work closely with his structural, electrical and mechanical engineers. Thus, the team approach would focus financial, architectural, and engineering talents on the problem of assessing the optimum height for an apartment or office tower.

INVESTMENT ANALYSIS The decision of what, where, and when to build is, of course, the basic decision determined either by a formal feasibility study or by a cruder, informal substitute. But the financial analyst's work is far from completed with his prediction that the project will be profitable. Once it has been determined that the project will meet the investor's criteria for a successful investment — i.e., yield a return equal to prevailing bank interest rates plus an additional percentage to compensate for additional risk — many problems remain in designing the precise transaction to suit the personal needs of the owner.

FEASIBILITY STUDY, ILLUSTRATIVE EXAMPLE

This example continues the Appraisal. Assume same basic data, plus

(5) Mortgage loan @ 75 percent appraised value, @ ten percent interest, for 25 yrs.

(6) Minimum acceptable return on cash investment=20 percent.

Step 1: Compute mortgage loan=0.75 x $1.2 million
$$=\$900,000$$

Step 2: Compute annual debt service payment=0.1091 x $900,000
$$=\$100,000 \text{ (approx.)}$$

Step 3: Compute profit before taxes =Net income—Debt service
$$=\$120,000-\$100,000$$
$$=\$20,000$$

Sept 4: For 20 percent minimum return on investment, compute the maximum cash investment allowable.
$$\text{Max. cash investment} =\frac{\$20,000}{0.20}$$
$$=\$100,000$$

Step 5: Compute max. project cost, or budget

$$\text{Max. project cost} =\text{Mortgage loan} +\text{Cash investment}$$
$$=\$900,000+\$100,000$$
$$=\$1 \text{ million}$$

Feasibility may be computed another way: by comparing maximum cash available from the developer with the estimated project cost and then checking for adequate before-tax return on cash investment. Assume the following data:

(1) Max. cash investment =$80,000

(2) Estimated total project cost =$1,020,000

Step 4A: Compute required cash investment

$$\text{Req'd cash investment} =\$1,020,000-\$900,000$$
$$=\$120,000 \text{ (50 percent higher than assumed limit)}$$

Step 5A: Compute pre-tax return $= \dfrac{\text{Pre-tax profit}}{\text{Cash investment}}$

$$= \frac{\$20,000}{\$120,000}$$

$$=17 \text{ percent}$$

INVESTMENT ANALYSIS, ILLUSTRATIVE EXAMPLE

FIRST-YEAR DEPRECIATION

Data: New residential apartment building.

(1) Total project cost $=$$130,000 ($30,000 for land plus $100,00 for building)

(2) Net operating income (gross income—total operating expenses) $=$$10,000/ year

(3) First-year mortgage interest payment $=$$7,000 (on 25-year loan at 9% interest, debt constant $=$0.1018)

(4) Developer's cash investment $=$$20,000

(5) Double-declining balance (200 percent) accelerated depreciation; 40-year useful life (per U.S. Treasury guidelines); total depreciable value $=$$100,000 (no salvage value)

Step 1: Compute first-year depreciation $=\dfrac{2 \times \$100,000}{40}=\$5,000$

Step 2: Compute net return $=$Net operating income—Interest payment
$=$$10,000—$7,000 $=$$3,000

Step 3: Compute taxable income $=$Net return—Depreciation
$=$$3,000—$5,000 $=$ —$2,000 (a $2,000 tax loss)

For a developer in the 50 percent tax bracket, the $2,000 tax loss has a twofold result:

(1) It makes the $3,000 net return tax free

(2) It subtracts $2,000 in taxable income from other income

Thus for a man in the 50 percent tax bracket, real estate depreciation can raise a 15 percent net return into an equivalent 40 percent return on ordinary income. To equal the $4,000 after-tax profit earned from the foregoing hypothetical project, ($3,000 tax-free income plus $1,000 in tax reduction), a man in the 50 percent tax bracket would have to earn $8,000 in ordinary income on the same $20,000 cash investment.

SALE WHEN TAX SHELTER EXPIRES

Needing cash for other ventures, the developer may decide to sell when the depreciation deduction fails to shelter his entire taxable income from the project. In this instance, the seventh year is the last in which total taxable income is sheltered, as shown in the tax shelter chart. (This problem can be solved mathematically.)

CAPITAL GAIN BENEFIT

Assume that the developer sells after the seventh year of ownership. Total selling price (land+building) =$140,000.

Step 1: Compute total deducted depreciation, S, by adding seven annual deductions or use following formula:

$$S = D_t [1-(1-c)^x]$$

in which D_t=Total depreciable value=$100,000

x=No. of years=7

c=Depreciation factor=$\dfrac{2}{n}$=.05

(n=years of useful life=40)

$$S = \$100,000 [1-0.95^7]$$
$$= \$100,000 [1-0.70]$$
$$= \$30,000$$

Step 2: Compute capital gain =Sale price—(Original cost—S)
$$= \$140,000 - (\$130,000 - \$30,000)$$
$$= \$40,000$$

Under the Tax Reform Act of 1969, with its "depreciation-recapture" provision, residential property is now taxed as follows: For property held less than 100 months, all excess depreciation is taxed as ordinary income.

Step 3: Excess depreciation $= \$30,000 - \dfrac{7 \times 100,000}{40}$
$$= \$30,000 - \$17,500$$
$$= \$12,500$$

Since the property was held seven years (84 months), none of the excess depreciation (i.e., that over the straight-line depreciation) qualifies for capital-gain treatment. Of the total $40,000 capital gain, $12,500 is excluded from capital-gain treatment. For an individual in the 50 percent tax bracket, the saving would amount to 25 percent of $27,500 =$6,900.

TAX SHELTER DECAY

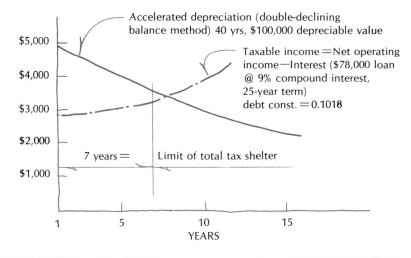

The name of the financial game is *leverage,* an apt metaphor conjuring up a vision of large profits pried up with small cash investments. By lengthening the "lever arm" of his loan with respect to his investment — the *debt-equity* ratio — the developer raises his profit percentages. If a project yields a $10,000 net annual return on a $10,000 cash investment, it has 10 times the leverage of a $10,000 return on a $100,000 cash investment. On mortgaged-out—100-percent financed—projects with relatively light incidental expenses paid from his own pocket, the developer may realize extremely high profit. Maximizing the developer's financial leverage is thus a basic goal of good economic design.

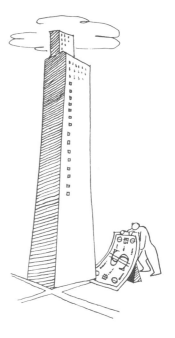

A classic job of successful leveraging, described in William Zeckendorf's autobiography, converted a $5000 cash investment into an $83,000 profit. This 1660-percent return on invested equity resulted from a complex, three-cornered 1940 real estate transaction. The story starts with the foreclosure of a $148,000 mortgage on an architect-developer's renovation project, converting a fashionable East Side Manhattan mansion into an apartment. (The architect lost the project because he badly underestimated the construction costs.) The rest of the story illustrates the importance of business acumen in real estate transactions.

Here's how Zeckendorf exploited the leveraging principle:

He bought the renovated apartment house with a $72,000 cash offer, obviously far below the actual economic value of the property, but acceptable to the Depression-minded mortgagee-owner, who wanted to avoid the risk of another foreclosure proceeding. With a $105,000, 4-percent interest mortgage, Zeckendorf put up only $5000 in cash and paid off the owner with the mortgage loan, immediately pocketing a $33,000 profit.

Zeckendorf then sold the apartment to a wealthy tenant, who was paying $6000 a year in rent, for $150,000. With the other apartments paying sufficient rent to carry the mortgage payments, maintenance, and other operating expenses, the new tenant-owner was home rent free. Zeckendorf turned over the $105,000 mortgage, and took $20,000 in cash plus two houses, which he later sold for $15,000 each, thus netting an additional $50,000 for a total profit of $83,000.

From this highly profitable transaction, Zeckendorf drew the following moral:

". . . By providing a *cash market* for Blumenthal (the foreclosing mortgagee), by doing this through the device of providing a 'safe, high value' *mortgage market* for the Troy Bank (the new mortgagee), and *then* by providing Mrs. Clews-Spencer (the final owner) with a self-liquidating income property, I had provided a needed service to three quite separate individuals, to everyone's immediate advantage and at a handsome profit."

Tax advantages are secondary, but vitally important, investment factors. No matter how extensive, a tax advantage cannot turn a losing deal into a winner. But real estate offers several unique tax advantages that weigh heavily in economic design. These tax factors may shape the final form of the financing deal.

Depreciation is, by far, the biggest of these unique tax advantages. In the early years of a project, it can reduce income taxes to zero, and even produce a paper loss shielding other income from taxation. Unlike other depreciable items — i.e., automobiles, office equipment, production machinery, etc. — real estate often appreciates in its early years when its depreciation is highest. Moreover, it offers another advantage: Even though the developer may have only a small equity invested in a property, he can depreciate the full value of a building. In short, he can eat his financial cake and have it, too.

Buildings and other physical improvements can be depreciated. (For obvious reasons, land is not depreciable.) Project cost, not appraised value, serves as the basis for depreciation. For a property that has been sold, this cost is readily established. A new project, however, presents a more complex problem. Expenses incurred and immediately written off must be deducted (along with land cost) from direct and indirect costs to determine depreciable value.

Depreciation can be computed as *straight-line* or *accelerated*. Straight-line depreciation, so called because it appears as a straight horizontal line when plotted against a time abscissa, is computed by merely dividing depreciable value by the assumed useful life, in years. Straight-line depreciation remains at a constant value throughout the project's entire useful life under one owner. (At each change of ownership, however, the depreciation game starts anew.)

Accelerated depreciation can be computed in one of two basic ways. In the *declining-balance* method, the first year's depreciation is computed as a multiple — 2.0, 1.5, or 1.25 — of the straight-line depreciation. The depreciable amount steadily decreases through the years, as the product of the multiple and the undepreciated balance decreases (see depreciation chart). In the declining-balance method, salvage value can always be assumed as zero, since there is always a depreciable balance remaining after the last year's depreciation.

In the second method of accelerated depreciation, the *sum-of-the-digits* method, the annual depreciation is computed as a perennially diminishing fraction of the total depreciable value. Its results are similar to those yielded by the double-declining (200 percent) balance method.

Benefits from accelerated depreciation and the various declining-balance method factors were slightly reduced by the Tax Reform Act of 1969. As one example, the double-declining balance method (using a factor of 2.0) is now limited to *new* residential property. Checking on the permissible current depreciable formulas, the useful lives, and the latest tax regulations governing various types of real property, new and used, is the lawyer's, tax accountant's or realtor's job.

Componentization is a complicated, secondary method of accelerating depreciation, for new buildings only. HVAC, plumbing, lighting-ceiling and several other building subsystems normally have shorter useful lives than the structure. Thus the owner can divide the building cost into subsystem costs, or component costs, and depreciate them separately, using shorter useful lives for the less durable subsystems. If, for example, the HVAC accounted for 20 percent of total construction cost, a useful life of 15 years, roughly one-third the useful life of most new buildings, would appreciably accelerate the depreciation for the total building.

Capital gains tax rates are another major benefit of real estate tax investment. Though real property can often be depreciated at accelerated rates during its early years, its actual value often appreciates during these years, mainly because of the skyrocketing price of land and construction throughout the United States. Thus the favored capital gains rates, which reduce the rates payable on sale profits for property held longer than six months, can mean great savings for investors. For those in high tax brackets, capital gains treatment can cut taxes from one-half to nearly two-thirds (see the investment analysis chart).

The owner may avoid any capital gain tax if he exchanges his property or invests the proceeds from the sale in a larger purchase in certain kinds of real estate. He may exploit another tax law to avert the immediate impact of a large capital gain, through an installment sale. A sale qualifies for installment-sale tax treatment if the owner receives less than a 30-percent payment on the sale price in the year of the sale. There is no benefit for a developer perennially in the 50-percent or higher bracket, since the maximum tax on a fully qualified capital gain is 25 percent of the total (or 50 percent of the half of the capital gain that is taxed). But a developer or equity-shareholder in a lower tax bracket may benefit from spreading the payments for a real estate sale into four or five equal payments. For advice on these matters, the architect needs a tax attorney or accountant.

The architect should investigate rigorously the project's financial prospects, possibly through his own independent investment analysis, to make sure that his prospective return is commensurate with the risks.

A basic problem confronting a developer with the luxury of choice is whether to seek maximum tax shelter for his project's net income (see tax shelter decay), or whether to sacrifice this tax shelter in return for greater financing and a consequent reduction — perhaps even total elimination — of equity investment. A developer lacking sufficient cash often has no choice. If he cannot assemble sufficient cash investors to put up the difference between a 75-percent mortgage loan (based on appraised value, not actual cost) and total project cost, then he often must accept the conditions attending loss of ownership. But even the developer with a choice may voluntarily surrender the tax-depreciation advantage. Consider a residential property in which the depreciation tax shelter yields the developer an average $2000 or so additional after-tax income on a $20,000 equity investment for the first eight years. Despite this apparently irresistible return, an aggressive developer might decide to sacrifice the depreciation tax shelter in return for 100-percent financing.

COMPARISON OF STRAIGHT-LINE VS. ACCELERATED DEPRECIATION

(200%, 150% DECLINING-BALANCE METHODS)

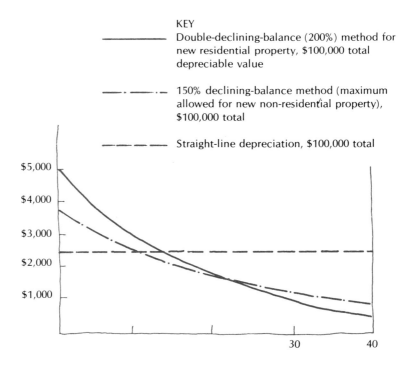

KEY

——————— Double-declining-balance (200%) method for new residential property, $100,000 total depreciable value

—·———·—· 150% declining-balance method (maximum allowed for new non-residential property), $100,000 total

— — — — Straight-line depreciation, $100,000 total

$$D_x = cD_T (1-c)^{x-1}$$

in which D_x = Depreciation for year x

D_T = Total depreciable value

c = Depreciation factor, $= \dfrac{2}{n}$ for double-declining balance method

$c = \dfrac{1.5}{n}$ for 150% method

n = Useful life (yrs.)

The developer could increase his financial leverage (net return-cash investment ratio) through a sale-leaseback deal that would reduce his cash investment to zero. In that event, the depreciation tax benefit would, of course, go to the lessor-owner. Nonetheless, for the aggressive developer who needs cash to exploit other opportunities, the sale-leaseback and variations on its basic theme have become increasingly popular. Discussion of all factors entering such a decision is beyond the scope of this book. It lies in the province of the financial analyst, another of the basic indispensable members of the development team. There are cash flow and taxable income checklists and a professional building's economic model at the end of this chapter.

Several words to the wise are appropriate at this point. First, a repetition of the familiar warning against an architect inexperienced in these matters to attempt a do-it-yourself approach. The choice of investment vehicle could hinge on minor tax revisions made almost annually. As one such example, the Tax Reform Act of 1969 removed part of the tax-shelter benefit by levying additional taxes against financial benefits not subjected to ordinary or capital-gain tax. Dealing with such matters is the work of professional tax attorneys and accountants, not amateurs.

A second thought concerns the need for safety factors in economic design. A structural engineer designs a floor beam with reserve strength, roughly twice that required to carry total design load. Market analysts, financial analysts, and cost consultants should think in similar terms in their approach to economic design. Allow for contingencies—higher than anticipated vacancy rates, construction costs somewhat higher than estimated, a construction time longer than scheduled. An ill-timed labor strike of even brief duration could ruin an otherwise economically sound project, lengthening the time and consequent cost of the construction loan, jeopardizing the permanent loan commitment, postponing the start of cash flow, raising construction costs, and generally playing havoc.

A last warning admonishes the reader not to underestimate the importance of formal economic analysis. If, for example, a developer decides to go ahead with a project rejected by the feasibility study, there are only two possible explanations: the developer rejects the implied premises of the study (i.e., the developer will accept more risk or less profit than the study implies), or he rejects the data used in the study. Economic design cannot be made perfectly scientific, but neither, for that matter, can structural design. The diffidence of some developers in formal economic analysis is like the rejection of critical path scheduling by some tradition-bound contractors. The neophyte developer is well advised to use all available technical tools, experience, and judgment. The validity of a feasibility study or investment analysis obviously depends on the accuracy of the input data. If, on the basis of his own experience or judgment, a developer decides to change the data in a study, he should make this decision explicit. In short, he should try to minimize the role of hunches, intuition, sixth senses, and blind guess. The risks in development work are difficult enough without deliberately clouding them with romanticism.

TRIAL ECONOMIC MODEL
HOSPITAL & MEDICAL PROFESSIONAL BUILDING

OUTLINE SUMMARY OF SALIENT INFORMATION

PROJECTION OF DIRECT & INDIRECT COSTS

1. Land — Note: The land will be on a net ground lease .... $	
2. Soil Tests & Hydraulic Analysis	1,500
3. Survey, Topographic Study, Traffic & Civil Engineering	3,500
4. Construction Costs	
a. Hospital — 156 Beds, 70,710 S.F. ($38.00PSF)	2,687,000
b. Medical Professional Building — 42,800 S.F. ($23.00PSF) .	985,000
c. Garage — 340 Cars, 118,000 S.F. ($5.00PSF)	531,000
5. Architectural & Engineering	262,000
6. Professional Fees (Interiors, Landscape, Tenant Layouts) ...	35,000
7. Legal	35,000
8. Project Overhead — 1.5% Approx.	90,000
9. Taxes During Construction	110,000
10. Insurance During Construction	1,000
11. Mortgage Loan Fee (1.5%)	90,000
12. Interim Loan Fee	60,000
13. Interim Interest ($6 Million, @ 9%, 12 Mos.)	540,000
14. Gap Loan Expense ($250,000 @ 9%, 1 Yr.)	22,500
15. Feasibility Study	7,500
16. Appraisal	5,000
17. Preparation of Loan Presentation	5,000
18. Furniture, Fixtures & Equipment	
a. Hospital	625,000
b. Medical Professional Building	10,000
c. Garage	10,000
19. Final Survey	740
20. Land Rental During Construction	125,000
21. Landscape	—
22. Permits, Licenses & Fees	4,500
23. Title & Closing	18,000
24. Contingencies	180,000
Total Direct & Indirect Costs	$6,444,240
Say	$6,450,000

INITIAL EQUITY AND WORKING CAPITAL REQUIREMENTS

Total Direct & Indirect Costs	First Mortgage	Equipment Mortgage	Working Capital Loan	Initial Equity
$6,450,000	*$5,650,000	$625,000	N.A.	$175,000

Note: The first mortgage amounts include a pro rata portion of the garage costs.

*Medical Professional Building 42,800 S.F. @ $30,000 PSF $1,282,000
Hospital — 156 Beds @ $28,000/Bed 4,368,000
Equipment — $625,000 — 5 Yrs. @ 7½% 625,000
$6,275,000
Initial Equity 175,000
$6,450,000

FIRST MORTGAGE REQUIREMENT

Amount	Term	Rate	Constant	Annual Payment
$5,650,000	27 Yrs.	9¾%	10.52	$594,380

EQUIPMENT MORTGAGE

Base	Amount	Term	Rate	Constant	Annual Payment
1.	$625,000	5	7½%	24.05	$150,312.50

PROJECTION OF INTEREST EXPENSE (FIRST MORTGAGE)

Year	Base 1	Year	Base 1
1.	$ 548,877	14.	$ 433,581
2.	544,237	15.	417,183
3.	539,123	16.	399,113
4.	533,489	17.	379,201
5.	527,279	18.	357,257
6.	520,436	19.	333,076
7.	512,896	20.	306,429
8.	504,586	21.	277,065
9.	495,429	22.	244,706
10.	485,338	23.	209,047
11.	474,219	24.	169,752
12.	461,965	25.	126,449
13.	448,462		$10,249,195

PROJECTION OF AVAILABLE INTEREST EXPENSE (EQUIPMENT MORTGAGE)

Year	Base 1	Year	Base 1
1	$ 43,244	4	16,322
2	34,932	5	5,920
3	25,975		$126,393

PROJECTION OF AVAILABLE DEPRECIATION

Total Direct & Indirect Costs $6,450,000
  Less Construction Phase Expenses 797,500
    Total Depreciable Assets $5,652,500

Notes:
  1. Assign 85% of T.D.A. to *Structures* on a 40-year economic life, Straight-Line Method.
  2. Assign 15% of T.D.A. to *Equipment* on a 20-year economic life, Straight-Line Method.

PROJECTED ANNUAL ALLOWABLE DEPRECIATION

Year	Structures	Equipment	Total
1.	$ 120,116	$ 42,394	$ 162,510
2.	120,116	42,394	162,510
3.	120,116	42,394	162,510
4.	120,116	42,394	162,510
5.	120,116	42,394	162,510
6.	120,116	42,394	162,510
7.	120,116	42,394	162,510
8.	120,116	42,394	162,510
9.	120,116	42,394	162,510
10.	120,116	42,394	162,510
11.	120,116	42,394	162,510
12.	120,116	42,394	162,510
13.	120,116	42,394	162,510
14.	120,116	42,394	162,510
15.	120,116	42,394	162,510
16.	120,116	42,394	162,510
17.	120,116	42,394	162,510
18.	120,116	42,394	162,510
19.	120,116	42,394	162,510
20.	120,116	42,389	162,505
21.	120,116	—	120,116
22.	120,116	—	120,116
23.	120,116	—	120,116
24.	120,116	—	120,116
25.	120,116	—	120,116
Total	$3,002,900	$847,875	$3,850,775

PRO FORMA PROJECTION OF ANTICIPATED INCOME AND EXPENSE

Income (Stabilized)

A. Hospital (156 Beds, 80% Occupancy, 45,552 Patient Days)
   Estimated Gross Earnings:

Routine Services	$2,869,000
Special Services	2,697,000
Other	171,000
	$5,737,000
Less Bad Debts & Contractual Allowance	258,000
Adjusted Gross Earnings	$5,479,000

B. Medical Professional Building

Doctors' Suites — 30,000 S.F. @ $7.20 PSF	$ 216,000
Retail/Commercial — 6,400 S.F. @ $6.00 PSF	38,400
Gross Income	$ 254,400
Adjust 5% for Vacancy	12,720
Adjusted Gross Income	$ 241,680

C. Garage

Income — 340 Spaces @ $240/Yr. Ea.	$ 81,600
Adjust 5% for Vacancy	4,080
	$ 77,520
Total Income All Sources	$5,798,200

Expense (Stabilized)

A. Hospital

Administrative	$ 428,000
Dietary	387,000
Household & Property	344,000
Professional Care of Patients	3,001,000
Management Fee	164,000
Total Expense	$4,324,000

B. Medical Professional Building

36,400 S.F. @ $2.00 PSF	$ 72,800
C. Garage (40% of $81,600)	32,640
D. Land Rent	50,000
Total Expense All Sources	$4,479,440

Net Income Before Debt Service

Income	$5,798,200
Expense	4,479,440
Net Income (Before Debt Service)	$1,318,760

CASH FLOW PROJECTION

Income		$5,798,200
Debt Service — First Mortgage	$ 594,380	
Debt Service — Equipment	150,312	
Expense	4,479,440	
Expense Total		5,224,132
Cash Flow		$ 574,068

INVESTMENT ANALYSIS

Year	Taxable Income/Loss	Cash Flow	F.I.T. 50%	Tax Loss Offset	Net Consequence	% Return on Equity
0	($ 797,500)	$ —0—	$ —0—	$398,750	$ 398,750	227.9
1	564,129	574,068	282,065	—0—	392,003	166.9
2	577,081	574,068	288,541	—0—	285,527	163.2
3	591,152	574,068	295,576	—0—	278,492	159.1
4	606,439	574,068	303,219	—0—	270,849	154.8
5	623,051	574,237	311,525	—0—	262,712	150.1
6	635,814	724,380	317,907	—0—	406,473	232.3
7	643,354	724,380	321,677	—0—	402,703	230.1
8	651,664	724,380	325,832	—0—	398,548	227.7
9	660,821	724,380	330,411	—0—	393,969	225.1
10	670,912	724,380	335,456	—0—	388,924	222.2
11	682,031	724,380	341,015	—0—	383,365	219.1
12	694,285	724,380	347,143	—0—	377,237	215.6
13	707,788	724,380	353,894	—0—	370,486	211.7
14	722,669	724,380	361,334	—0—	363,046	207.5
15	739,067	724,380	369,535	—0—	354,845	202.8
16	757,137	724,380	378,568	—0—	345,812	197.6
17	777,049	724,380	388,524	—0—	335,856	191.9
18	798,993	724,380	399,497	—0—	324,883	185.6
19	823,174	724,380	411,587	—0—	312,793	178.7
20	849,826	724,380	424,913	—0—	299,467	171.1
21	921,579	724,380	460,790	—0—	263,590	150.6
22	953,938	724,380	476,969	—0—	247,411	141.4
23	989,597	724,380	494,799	—0—	229,581	131.2
24	1,028,892	724,380	514,446	—0—	209,934	120.0
25	1,072,195	724,380	536,097	—0—	188,283	107.6
Total	$17,945,137	$17,358,109	$9,371,320	$398,750	$8,385,539	N.A.

PROJECT-COST PROJECTION CHECKLIST

PROJECTION OF DIRECT AND INDIRECT COSTS

A. Land

    1. Survey   _____

    2. Soil Test   _____

    3. Abstract or Title Policy   _____

    4. Legal   _____

    5. Appraisal   _____

    6. Real Estate Fees   _____

    7. Interest (Holding Costs)   _____

    8. Demolition Costs   _____

    9. Feasibility Studies   _____

    10. Taxes (Holding Costs)   _____

                   TOTAL LAND   _____

B. Site Preparation   _____

C. Architectural & Engineering   _____

D. Contractor's Cost Estimate   _____

    1. Buildings   _____

    2. Paving & Site Work   _____

E. Mortgage Loan Fee   _____

F. Interim Loan Fee   _____

G. Interim Interest   _____

H. Insurance During Construction   _____

I. Taxes During Construction   _____

J. Appraisal Fee   _____

K. Legal Fees   _____

L. Accounting   _____

M. Project Overhead   _____

N. Title Costs   _____

O. Survey   _____

P. Landscaping   _____

Q. Furniture, Furnishings, Fixtures & Supplies   _____

R. Professional Fees (Decorators—Mg't. Studies)   _____

S. Advertising & Public Relations   _____

T. Permits, Licenses & Fees   _____

U. Performance Bond   _____

            Total Direct & Indirect Costs: _____

OPERATING EXPENSES CHECKLIST

	Submitted	Stabilized
Taxes (Ad Valorem)		
Insurance		
Payroll		
Payroll Taxes & Fringe Benefits		
Utilities		
Water		
Gas		
Electricity		
Contract Maintenance		
Scavenger		
Elevator		
Pest Control		
Heat—Vent. & A/C		
Landscaping		
Painting & Decorating		
Advertising		
Legal		
Accounting		
Supplies		
Repairs & Replacements		
Equipment		
Furniture		
Building		
Reserves for Replacement		
(or actual cost of replacement)		
Carpets		
Drapes		
Equipment		
Appliances		
Etc.		
Management		
Parking		
Licenses, Permits, Fees & Business Taxes		
Contributions		
Subscriptions & Dues		
Contingencies		
Transportation & Travel		
Communications		
Entertainment		
Total:		

TAXABLE INCOME* CHECKLIST

Income (all sources) $ _____

    Less:
        Interest Payments $ _____

    Operating Expenses:
        Taxes $ _____

        Insurance   _____

        Maintenance   _____

        Utilities   _____

        Management   _____

        Etc.   _____

    Depreciation   _____ $ _____

              TAXABLE INCOME/OR LOSS $ _____

                                (   )

*Taxable income is that portion of income subject
to federal income tax after deductions for interest
payments, operating expense, and depreciation.

CASH FLOW* CHECK LIST

Income (all sources) $ _____

    Less:
        Debt service (Principal +Interest) $ _____

    Operating Expenses
        Taxes (real estate, etc.) $ _____

        Insurance   _____

        Maintenance   _____

        Utilities   _____

        Management   _____

        Etc.   _____   − $ _____

            CASH FLOW = $ _____

*Cash flow is total annual income minus all paid-out expenses

LAND ACQUISITION

"Financial terms of land acquisition can become the pivotal point in successful project generation and development."   — Kenneth Schnitzer

Along with economic design and negotiation of a financing commitment, land acquisition ranks as one of three major decision-stage tasks. Not only is it an indispensable step; it also is a highly critical step in the development schedule. It must be completed before the project's financing can be committed.

In land acquisition, as in every other big monetary transaction associated with project development, the developer attempts to minimize the cash investment to attain maximum financial leverage—the profit-cash investment ratio. In conformance with this leveraging principle, the developer seeks to delay cash outlay for land purchase as long as possible—or better yet, to avoid cash outlay. He may achieve these goals, in varying degree, through several available techniques:

▪ Deed-and-purchase-money mortgage ▪ Land contract ▪ Option on land purchase   ▪ Ground lease   ▪ Joint venture   ▪ Deferred acquisition ▪ Syndication.

In a deed-and-purchase-money mortgage, the developer buys the land in the same way that a homeowner buys a house. He makes a small down payment and then executes a purchase-money mortgage on the land, with the landowner acting as mortgagee. The purchase-money mortgage is, in effect, a mortgage loan, with the landowner-mortgagee holding the fore-closure right to reclaim his property if the land-buying developer defaults in his annual payments (interest plus principal). His foreclosure right to the land depends upon whether the mortgagee for subsequent development agrees to "subordinate" his first lien rights on the development. The land seller's retaining the first foreclosure right or subordinating his rights to the permanent mortgagee is always agreed upon in advance. The term of such land purchase-money mortgages generally runs about five years, with equal annual installments.

If an architect's financial interests in a given class of projects grow so large that he becomes a substantial market force, his financial interest in that field may subvert, or appear to subvert, his professional interests.

The land contract resembles the deed-and-purchase-money mortgage, with one major exception. In the deed-and-purchase-money mortgage, the title deed is transferred to the purchaser when he makes his initial down payment. In the land contract, the original owner retains title to the property until the developer completes his installment payments (including interest). Meanwhile, the developer gets full use of the land. If the project is built in stages, the developer probably will get the use of the land in stages. His payments must parallel his assumption of the land: when he takes over 50 percent of the land, his payments must total 50 percent of total purchase price.

Options on land enable the developer to tie up land with minimal cash outlay, possibly 5 percent to 10 percent of the ultimate cost. With land prices climbing faster than 10 percent a year over the past several decades, options usually contain a cost escalation clause. A series of successive options, set for annual expiration dates, will provide for the price per acre to rise from, say, $40,000 per acre in 1972 to $45,000 in 1973. Despite this disadvantage of escalating cost, however, the option admirably achieves the developer's normal goal of minimizing early cash investment.

Ground leases, negotiated with landowners who for various reasons may not want to sell, require no cash investment for land. At annual rents ranging several percentage points above or below 10 percent of the land's market value, the developer gets use of the land until the original lease expires (perhaps after 40 years), with renewal options usually for another 20 years. During the leasing periods, the developer owns the buildings. Their ultimate transfer to the landowner, when the final lease term expires sometime after the year 2000, leaves the developer unconcerned.

Joint venture participation by the landowner eliminates cash investment for land by giving him equity shares in the project. Suppose the land cost is $100,000 and total required cash investment is $250,000. If the landowner were willing, the developer might offer him a 40-percent share in the

Both ground and air rights were utilized in Atlanta's Peachtree Center development — a $150 million office, mart, hotel, retail, and entertainment center. The aerial walkway links the two-million-square-foot Merchandise Mart to the Gas Light Tower. Peachtree Center's developers were Trammel Crow, of Dallas. John Portman & Associates designed the center. The 33-foot-high sculpture in the foreground was designed by Robert Halsmoortel.

project's profits in return for the landowner's contribution of the land. The landowner becomes a partner in a joint venture with the developer and other partners.

Deferred acquisition is a slightly more complicated technique designed to reduce the risks accompanying a lease or joint-venture land acquisition. This technique resembles an option without the cost. It postpones consummation of the land-acquisition transaction until the delivery stage, thus eliminating the costs and/or embarrassment attending a possible abandonment of the project during the decision or design stage. These deferred-action plans do, however, require a special kind of landowner—one who ultimately plans to develop his land rather than sell it.

Here's how a deferred-acquisition transaction works: An architect-developer finds a landowner with a valuable tract. He convinces the owner that the land should be developed. An attorney draws up a letter of agreement containing the following provisions:

▪ The owner's agreement to hold the land for an interval ranging from six to 12 months while the architect does market and feasibility studies, and possibly a highest-and-best-use study.

▪ On completion of the first step, the owner's further agreement to hold the land for a longer period, to allow the architect time to complete schematic design and to arrange for financing.

▪ The owner agrees to hold the land for a still longer period, contingent on the architect-developer's obtaining a financing commitment, to allow the architect time to design the project and to arrange interim financing.

▪ The owner finally agrees either to lease the land or to take a joint venturer's equity share.

Initiation of the landowner's payments depends on whether he leases his land or joins the development team as an equity sharer. If he leases the land, his payments start with the interim loan payments. If he takes an equity share, his return is probably deferred until the project is completed and cash starts to flow.

Before he starts any work, an architect-developer contemplating a deferred-acquisition transaction should get the landowner's signature on a tightly drafted letter of agreement. As news spreads that the particular land tract is open to development, the owner may become a target for other proposals. If the owner accepts one of these counterproposals, he can leave the architect holding a bag of wasted economic studies and schematic drawings. The architect's only legal protection from such a fate is a letter of agreement.

Syndication is still another technique for tying up land. A real estate syndicate is an investment group organized by a broker-syndicator to split large investments into relatively small shares, thus allowing small investors to compete with big investors in land speculation. A syndicate can be a general partnership, a corporation, a real estate investment trust, a joint venture, or a joint stock company. The most popular form, however, is the limited partnership. In this form of syndicate, the broker-syndicator is a general partner sharing in the partnership management. The small investors

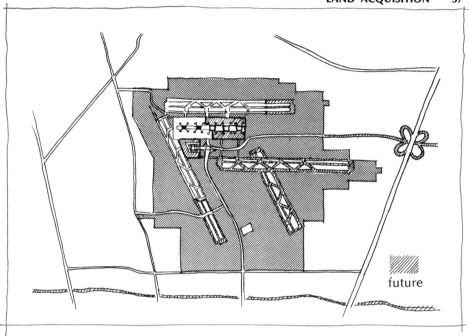

future

Houston's Intercontinental Airport influenced land values north of the city. Land which sold for $600 per acre in 1961 was selling for up to $100,000 per acre in 1972. Airport architects were Golemon & Rolfe, and Pierce, Goodwin & Flanagan, of Houston.

are the limited partners, with no voice in management. A developer who uses a real estate syndicate to acquire land needs an attorney to investigate the many complex tax and organizational problems.

When the time comes for development, the land-owning partners probably will find it advantageous to dissolve the land syndicate and form a new joint venture. Creation of a new development organization and dissolution of the land syndicate offer several advantages over the alternative policy of carrying out development work under the original land syndicate. Legally, a joint venture, and later, possibly a corporation, offer more suitable organizational vehicles for a development project. Dissolution of the land-owning syndicate also allows some members, if they so desire, to drop out of the development project. The organizational change also enables the project to start with a fresh accounting ledger.

When he selects a site, the market analyst investigates transportation—current and planned access to airports, general highway, and mass transit.

Landowners often sell at below-market prices during such organizational change from landholding to development. As joint venturers in a new development project, they defer capital gain tax far into the future, and possibly forever, by reducing, or eliminating, their land-sale profits. And since they are, in effect, selling the land to themselves, they lose absolutely nothing in available cash.

Typical of the projects spawned by these land syndication deals is a planned Long Island industrial park. Several years ago, six land investors, including Robert W. Jones, AIA, a vice president of Shreve, Lamb & Harmon Associates, of New York City, formed a limited partnership to buy a 120-acre site near MacArthur Field in Smithtown, Long Island. In early 1972, the syndicate was negotiating a land sale to a newly formed development corporation. Half the $2.5-million sale price will be paid in cash, with the remaining half distributed as equity capital (corporation stock) to launch the development project. SL&H, the project planner-architect, may take part of its fee as stock in the development corporation.

Big-time entrepreneurs perennially in the development business often acquire land several years—even a decade—before they build. Strategically located land is usually an excellent investment in its own right. In some urban areas, land prices have multiplied 10 or 20 times within the past decade.

Fluctuating land values around the new Houston Intercontinental Airport illustrate the major forces at work. In 1961, before the approval of a bond issue for financing the new airport construction, land sold for $600 per acre. By 1972, the same land was selling for up to $100,000 an acre depending upon accessibility to streets and other improvements. Anticipating the local boom expected from the airport, developers are planning a host of projects—industrial plants, motels, office buildings, shopping centers, and apartments. Yet despite the general rise in land values, the values of some tracts have fluctuated with changing plans for freeway interchanges or orientation of the airport runways. Some recent purchasers may have been burned financially by the relocation of a planned freeway interchange. Land immediately adjoining the airport under the landing pattern dropped in value once it became known that this land would be most heavily assaulted by the jet noise. Architects drawing on their daily experience dealing with such basic planning factors are well qualified to project future land values for their clients or for themselves.

As urban growth continues its relentless trend in and around our major metropolises, the scarcity of land has forced the creation of synthetic land —in the reclamation of swamps, meadowland, and refuse dumps, and in the use of air rights over existing structures—highways, railroad tracks, or even rivers. The environmental consequences of air-rights buildings, especially those subjected to the intense air pollution from congested streets and highways, require careful study. But the intensifying economic pressures in central business districts where land values are measured in hundreds of dollars per square foot cannot be ignored. One of the earliest air-rights deals was an 80-year lease of a one-acre air space over New York City's Grand Central Terminal for $1-million-plus a year. This air space encloses the 59-story, 2.4-million-square-foot Pan Am Building, which reigned briefly during the 1960s as the world's largest office building. Most air-rights transactions are negotiated on a similar basis—as a lease for an envelope of space above a specified elevation over an existing structure. Sometimes, as in the case of the Pan Am Building and in several other air rights buildings erected over Penn Central railroad tracks, there are provisions for the air-rights structures' foundations and lower-story columns to occupy land below the leased air space. But in some instances, the air-rights structures must bridge over the land below, thus necessitating some form of land acquisition on adjacent land.

Chicago is the site of an air rights project destined for an ultimately larger size than the Pan Am Building. As a principal in Ogden Development Corporation, co-developer along with Illinois Central Industries, and Charles Luckman Associates, the project architect, Charles Luckman, FAIA, has designed a $50-million office building as first stage of a huge 173-acre commercial complex using Illinois Central Railroad air rights. Such projects will predictably multiply in the future.

A knowledge of these varying techniques for tying up land is essential to an architect planning an entry into development work, either as a co-owning equity participant or as a consultant offering expanded decision-stage services to clients. For architects with limited capital, the various methods of reducing, postponing, or even avoiding cash outlay offer a means of participating in development enterprises that would otherwise be closed to them. On the other hand, for architects with substantial resources, land investment offers good investment prospects. Nothing else so certainly can assure the architect's status as a decision-maker in the development process as ownership, or co-ownership of the project site.

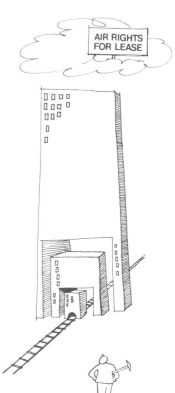

FINANCING A DEVELOPMENT PROJECT

"There is not a set amount of money . . . available for real estate development each year. All real estate ventures must compete on the open money market for available funds with other investment opportunities, such as stocks and bonds and equity investments."
                                                                    — Peter E. Pattison

Financing, along with location and market, is one of the basic factors that normally determine a project's success or failure. Often it is the deciding factor in the decision to build. Thus the mortgage banker's role—finding good sources of capital on favorable terms for the developer—is crucial.

Chronologically, financing is the last of three major tasks, following economic design and land acquisition, to be accomplished during the decision stage. With the commitment for a mortgage loan—or some alternative means of financing—a development project is ready to roll into the design and delivery stages.

All development financing techniques fall into one of two basic categories: *debt capital* or *equity capital*. Compared with the more familiar private corporation financing techniques, debt capital corresponds with corporate bonds, whereas equity capital corresponds with common stock. Debt capital usually is loaned at a fixed interest rate for a fixed term of years. Like corporate bondholders, these debt-capital lenders hold first claim on earnings or on proceeds obtained from a bankruptcy sale. Equity capital fills the financial gap between the borrowed funds and total project

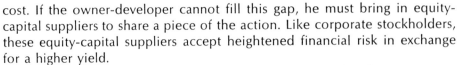

cost. If the owner-developer cannot fill this gap, he must bring in equity-capital suppliers to share a piece of the action. Like corporate stockholders, these equity-capital suppliers accept heightened financial risk in exchange for a higher yield.

In recent years, the conventional financing pattern has become enormously complicated because lenders, exploiting tight mortgage-money markets, increasingly have demanded equity participation as a condition for supplying debt capital. Many new financing techniques—sale-leaseback, sale-buyback and other innovations described later—have been developed in response to new demands. There are many new sources of equity capital—landsellers, architects, lawyers, doctors, realtors and other development team members, bankers, private corporations and institutional lenders.

Despite the rising incidence of equity-sharing techniques, however, debt capital remains the basis of project financing. A mortgage banker normally serves as broker between the developer (borrower) and investor (lender) in negotiating two basic loans:

■ A *mortgage* (or *permanent*) loan, delivered after project completion and usually repaid in monthly installments over a 20- or 30-year period.

■ A *construction* (or *interim*) loan, used to pay construction costs, professional fees, land improvement costs, property taxes, interest charges, and other project costs that fall due prior to or during the construction period. This short-term debt is normally payable on or before project completion.

As the basis for the construction loan, the mortgage loan is the key to good financing. Because the mortgage lender pays off the construction lender's loan, the mortgage loan normally must be committed before the construction loan can be negotiated. Conversely, because the project is dependent upon the construction loan, the permanent mortgage lender often requires *confirmation* of the forthcoming construction loan. The mortgage loan also sets the limit for the construction loan.

The mortgage loan is secured by the familiar first mortgage. A first mortgage gives the lender foreclosure rights—i.e., the right to sell the project to recover his investment from the proceeds if the developer defaults on his debt-service payments.

State law generally sets 75 percent of total appraised value as the limit of the mortgage loan advanced by life insurance companies, and other prudent lenders generally observe this limit.

Once the permanent loan is committed, the full amount often is advanced only after the confirmation of a specified occupancy level, particularly on speculative projects. The unqualified portion (that portion advanced regardless of the occupancy level requirement) of the mortgage loan becomes the limit of the construction loan, since the construction lender depends on the mortgage lender for repayment.

The methods of repaying construction vs. mortgage loans differ. On the short-term construction loan, the borrower pays simple interest, usually in monthly installments until the loan term expires, when he must repay the entire loaned amount, or principal. On the long-term mortgage loan, however, the borrower must amortize the loan—i.e., he must repay the debt in regular installments that include both interest and principal.

OBTAINING THE FINANCING COMMITMENT The search for project financing may proceed something like a treasure hunt. The mortgage banker operates like a detective, steadily narrowing the focus of his inquiries to identify the eventual lender. The financing search generally starts with an effort to obtain a conventional mortgage loan. An alternative financing technique ultimately may be chosen as the financing vehicle. But regardless of its final details, a successful search generally proceeds through the following steps:

■ Circulation among prospective lenders of a financial package describing the project.

■ Submission of a preliminary formal application to a prospective lender, whose interest may have been aroused by the financial package brochure.

■ Payment of 1- to 2-percent standby fee to the prospective lender.

■ Submission of a final, revised mortgage loan application letter, made out in conformance with the lender's instructions, for final approval by the lending institution's loan committee.

■ Signing of the mortgage loan commitment by the developer-borrower and lender.

Once the development team has negotiated the mortgage loan commitment, the financing chronology resumes with the following steps:

■ Negotiation of a construction loan (before construction starts).

■ Payment of the construction loan by the mortgage lender (when construction is completed).

■ Repayment of the developer's deposit—minus attorney's fees, etc.—by the lender.

Preparing the mortgage package. The mortgage banker, assisted by the architect, prepares a financing package including the owner's financial statements and copies of advance leasing agreements negotiated with lessees. As the predominant part of this package, a brochure should contain the basic information required by the lender to evaluate the project's investment prospects, specifically the following:

■ Brief introductory description of the project, the development team and the economic prospects.

■ Design concept, depicted with schematic drawings—renderings, site plan, typical floor plans.

■ Description of known lessees.

■ Summary market and feasibility studies.

■ Program of services and facilities.

■ Technical description of project—sitework, building materials, mechanical equipment, airconditioning, elevators, etc.

■ Site description—photograph, map.

■ Comparable land values in area, sale prices of nearby tracts.

■ Developer's pro forma (investment projection), including summary of total (direct plus indirect) costs, equity investment, first mortgage loan requirement, and cash flow (see page 44).

■ Development team members—qualifications and experience.

The formal application letter. After the preliminary survey of potential lenders has turned up a prospective lender, the mortgage banker submits a formal application letter accompanied by a 1- to 2-percent standby fee. This letter expands the basic information in the mortgage package. It would contain:

■ Data on owners or sponsors—including key personnel with their experience, credit reports from Dun & Bradstreet or similar sources, financial statements indicating source of funds for equity investment.

■ Project description, including location, acreage, net usable space, gross area and space use, usually illustrated with aerial photographs.

■ Plot plan, showing property dimensions, topography, streets, building restrictions, zoning, and limits on tenancies, with building outlines and parking areas outlined and noted with gross area and number of parking spaces.

■ Plans and elevations, notably typical floor plan, typical individual units and common or utility buildings.

■ Builder's experience, plus data relative to the construction completion date.

■ Consultants' listing—architects, engineers, planners, realtors, and other members of the development team, plus their experience.

■ Cost analysis: detailed project cost estimates (land, fees, construction costs, with itemized breakdown for materials and labor).

■ Leasing schedule, including estimated rent and leasing terms, nature of business or businesses, gross area, renewal clauses, and other data depending on nature of project.

■ Detailed accounting of estimated income and expenses, indicating amount of net income allocated to debt service.

■ Land acquisition data—copies of land contract, deed, or option to buy.

■ Competing facilities, to be shown on municipal map, with distance from proposed project, size, rent structure, tenant characteristics, parking, etc.

■ Market feasibility report, to be provided by independent analyst. (Though not always mandatory, this item should normally be included to aid the lender.)

Negotiating the mortgage loan. The most important terms negotiated between the lender, on one hand, and the mortgage banker and his developer, on the other, are: ■ interest rate, ■ mortgage term an ' ■ prepayment privileges and other especially negotiated conditions.

The interest rate is usually fixed, with debt-service charges set for constant annual payments. (Debt service is the total payment of interest plus

CONSTANT ANNUAL PERCENT
(DIVIDE BY 12 TO DETERMINE MONTHLY PAYMENT)

Interest Rate	20 YR	21 YR	22 YR	23 YR	24 YR	Interest Rate	25 YR	26 YR	27 YR
7	9.31	9.11	8.93	8.76	8.62	7	8.49	8.37	8.26
1/4	9.49	9.29	9.11	8.95	8.81	1/4	8.68	8.56	8.46
1/2	9.67	9.47	9.30	9.14	9.00	1/2	8.87	8.76	8.65
3/4	9.86	9.66	9.49	9.33	9.19	3/4	9.07	8.96	8.85
8	10.04	9.85	9.68	9.53	9.39	8	9.27	9.16	9.06
1/4	10.23	10.04	9.87	9.72	9.59	1/4	9.47	9.36	9.26
1/2	10.42	10.23	10.07	9.92	9.79	1/2	9.67	9.56	9.47
3/4	10.61	10.43	10.26	10.12	9.99	3/4	9.87	9.77	9.67
9	10.80	10.62	10.46	10.32	10.19	9	10.08	9.97	9.88
1/4	11.00	10.82	10.66	10.52	10.39	1/4	10.28	10.18	10.09
1/2	11.19	11.01	10.86	10.72	10.60	1/2	10.49	10.39	10.31
3/4	11.39	11.21	11.06	10.93	10.81	3/4	10.70	10.60	10.52
10	11.59	11.61	11.26	11.13	11.01	10	10.91	10.82	10.71

Debt-constant table shows annual debt service payments (interest *and* principal) as a percentage of total loan. To repay a loan of $100,000 at 10% interest, 20-year term, would require annual payments of $11,590 (=11.59% x $100,000).

Constant annual payment = $10,040

INTEREST

PRINCIPAL

$10,000

$5,000

0   5   10   15   20

Constant annual payments (paid monthly) on a $100,000 loan at 8%, 20-year term, total $10,040 a year. Total payments = 20x10,400 = $208,-000, with $108,000 total interest (cross-hatched area) and $100,000 total principal (white area).

principal required to repay the mortgage loan.) At constant annual debt-service payment, the interest part of the payment starts high and steadily declines through the years as the principal is amortized. Conversely, the principal payments start low and steadily rise. For a $1-million mortgage loan at 8.75-percent interest for 30-year term, the constant annual debt service is 9.45 percent of the principal, or $94,500. Such a loan is denoted in a financial prospectus as follows:

8.75%, 30 years, 9.45% C(constant)

Though traditionally fixed, interest rates sometimes vary. They can be set to change with prevailing mortgage rates, or even the prime interest rate (the rate charged by Federal Reserve member banks to prime customers).

From the developer's viewpoint, *prepayment privileges* are more appropriately called *prepayment penalties*. They are designed to reimburse the lender for a sudden loss of his investment if the developer, for one reason or another, decides to pay off his loan in the early years. Prepayment provisions levy penalties for early payoffs. In a mortgage contract with a 25-year term, the prepayment clause might allow no payment within five years, with a 3-percent penalty on the unpaid principal for prepayment within 10 years, declining by 0.5 percent annually from the 10th through the 14th year to a minimum of 1 percent after the 14th year. Thus if a borrower pays off a $500,000 balance remaining at the end of year 12, he would pay a penalty charge of 3 percent $-(2 \times 0.5$ percent$)=2$ percent, or $10,000. The theory behind prepayment penalties is that the lender is entitled to reimbursement of his start-up costs and should not be penalized for loss of income.

In addition to the basic negotiations for the mortgage loan, the mortgage loan may include:

■ Additional mortgage commitment fees.

■ Provisions (if any) for equity participation in profits.

■ Loan holdback clause (restricting the loan amount to a specific percentage of the full amount until a specified project occupancy is achieved).

■ Required form of documents.

■ Miscellaneous provisions concerning security, leases, closing date, escrow requirements, commitment termination conditions, liquidated damages and other legal details.

THE MORTGAGE BANKER   Speed in securing a mortgage commitment is vital in times of rapidly escalating construction costs. A good mortgage banker normally should secure a mortgage commitment within 45 days after receiving the required information from the developer. Costly delays of six months are not uncommon for mortgage bankers who lack contacts with major lenders actively seeking good investment opportunities. Thus merely through fast action, a competent mortgage banker can save a developer several times his commission.

In selecting a mortgage banker, the developer should note the distinction between a *mortgage banker* and a *mortgage broker*. A mortgage broker is essentially an independent operator who acts as middleman between borrower and lender in negotiating a loan. He is free to place loans with many lending institutions. In general, the advantage offered by a mortgage broker is fast action. On the other hand, a mortgage banker acting as correspondent or agent for a smaller number of lending institutions offers a broader range of services. A mortgage broker usually appraises the project, but offers no further servicing. A mortgage banker, however, acting in his fiduciary capacity, stays with the project. He coordinates the closing transaction. He makes sure that the borrower continues to pay taxes and that he carries and maintains insurance. He collects the debt-service payments and makes periodic physical inspections. If the developer defaults on his payments, the mortgage banker assists in the foreclosure proceedings. A borrower pays from 1 to 2 percent of the total mortgage loan to the banker or the broker after the loan is secured.

THE LENDER   A single lender sometimes supplies both construction and mortgage loans. Such consolidation promotes general efficiency in financing operations, offering mutual advantages both to the developer-borrower and to the investor-lender. For the developer, it may also reduce carrying charges—service fees and other miscellaneous expenses.

Despite the advantages of consolidating mortgage and construction loans, most development financing still is handled by two lenders, each with his own specialized expertise. Among the major sources of construction loans are commercial banks, savings and loan associations, real estate investment trusts, and other aggressive investing institutions seeking the highest effective interest rates. The more conservative mortgage lenders include insurance companies, pension funds, savings and loan associations, and corporations more interested in secure, long-term investment than the construction lenders. Interest rates for mortgage loans generally run about 1 percent to 2 percent below the construction loan rate.

The most important terms negotiated between the lender and the developer and his mortgage banker are the interest rate, mortgage term, prepayment privileges and other especially negotiated conditions.

Two practices often used in conjunction with construction loans raise their true interest rate, as differentiated from the nominal rate, above the normal 1- to 2-percent differential. Some commercial banks require the borrower to maintain a "compensating balance." If a borrower must leave on deposit a 10-percent compensating balance (without interest) for a loan at a nominal 10-percent interest rate, the true interest rate is 11.11 percent. Discounting is another means of raising true interest rate. In discounting, part of the face value of the loan, measured in "points" or percentage (1 point=1 percent), is deducted from the nominal loan value. If, for example, a $1-million construction loan paying a nominal 9 percent is discounted 3 points, the true interest rate is 9.56 percent. On both a discounted loan and a loan requiring a compensating balance, the borrower pays interest on money he never gets to use. But in discounting, the disadvantages are compounded. In addition to paying a higher true interest rate than the nominal rate, the borrower repays the lender a larger principal sum. On a loan discounted 3 points, the repaid principal is 3.09 percent greater than the actual principal: $100/(100-3)=1.0309$. A borrower should always calculate true interest rate in assessing the merits of a prospective loan and comparing it with alternatives.

HISTORICAL CHANGE—DEBT TO EQUITY FINANCING In recent years, the fixed-rate interest-paying mortgage loan has lost its former predominance as a development financing technique. As recently as the mid-1950s, the vast majority of the nation's large multi-million-dollar commercial projects were financed solely via conventional mortgage loans. Between 1955 and 1970, however, the number of commercial projects financed exclusively by fixed-interest mortgage loans dropped from about 95 percent to 35 percent, according to a prominent New York City mortgage banker.

Following World War II, debt capital was plentiful from institutional lenders content with modest returns. During the euphoric economic booms of the late 1940s and 1950s, sometimes aided by inflated appraisals, commercial developments were often mortgaged out—i.e., financed with mortgage loans covering 100 percent of project cost. Federally guaranteed loans on many multi-family housing projects often required only 10 percent or less cash investment by the developer. Equity funds often came from a single developer, or from a local syndicate operating on a small scale. During this era of easy financing, developers could ignore the more complex financial mechanisms used by larger publicly held corporations.

The end of the easy-money era, effected in the 1960s, stemmed from three basic factors:

■ Increasing project size ■ Accelerating inflation rates ■ The Tax Reform Act of 1969

The trend toward ever larger projects has strained the existing sources of debt and equity capital beyond their capacity. In the mid-1960s, the advent of new towns like Reston, Va.; Columbia, Md.; Clear Lake City, Tex.; and Park Forest South, Ill., stretched commercial development into a new dimension, requiring new sources of capital supplied largely by major American corporations—Gulf Oil, Westinghouse, Firestone, General Electric. Unlike smaller subdivision projects, large-scale new towns require "patience money"—the investment of millions in sewage systems, water supply, roads, and even golf courses—years before cash begins to flow. Large-scale regional shopping centers and planned-unit developments require far more patience money than their smaller scale predecessors. The difficulty in raising this front-end money partially offsets the long-term advantages of superior, larger scale planning that gradually is supplanting the smaller scale, more speculative projects characteristic of the recent past.

An even more pervasive force behind the upsurge in equity-sharing financing techniques is accelerated inflation. Over the past five or six years, inflation has critically eroded the financial integrity of long-term investments bearing low, fixed-rate interest charges. And the Tax Reform Act of 1969, by placing corporations on more equal footing with individual investors, has encouraged corporations to participate in real estate ventures.

The mutual benefits for both borrowers and lenders in the complex new equity-sharing financing techniques are manifold. For some developers, equity sharing is necessary to permit expansion where, for example, the success of a large shopping center or industrial park creates the need for additional development and the capital for financing it.

Investors' motives basically are similar to those of the developer. Life insurance companies and other institutional lenders are seeking a better hedge against inflation, to protect insurance and annuity contracts spanning one or more generations. Lenders, like developers, often seek tax shelter. Equity sharing is also a means of averting usury laws, which limit mortgage interest rates below market rates in some states. Still another investor motive for equity sharing is to reserve opportunities for a first option on future financing of regional shopping centers, industrial parks, apartment complexes, or other large-scale projects developed in stages. As part owner, the equity-sharing lender gains a voice in future development decisions.

OTHER DEVELOPMENT FINANCING TECHNIQUES The proliferating techniques currently used for financing real estate development projects present a bewildering complexity if they are simply listed and mechanically described. Grouping them into general classes brings order into the apparent chaos. In their most general aspect, they divide again into debt financing and equity financing.

As the major vehicles for debt financing, the previously discussed mortgage and construction loans form the basic financial framework to which the other financing techniques, both debt and equity, are attached. In one form or another, the mortgage loan is the structure supporting the entire financial edifice. Among the supplementary debt-financing techniques are gap and standby financing and secondary mortgages of various kinds.

The need for gap financing, or a standby commitment, arises when the unqualified portion of the mortgage loan, due on project completion, is less than the total amount needed to pay the developer's expenses during the construction period. A developer might, for example, need $1 million to pay his total construction costs while the mortgage lender's commitment calls for only $800,000 (80 percent of the total $1-million mortgage loan). The construction lender prudently limits his loan to the unqualified part of the mortgage loan, which is used to pay him. Thus the developer needs a short-term loan of $200,000 until the required 85-percent occupancy is achieved and the $200,000 permanent-loan balance is due. Time limits of six to 18 months are often set for attainment of required occupancy levels. The developer's failure to meet any time limit for required occupancy nullifies the lender's responsibility to advance further funds.

As the most straightforward approach to this problem, the developer can negotiate a separate gap-loan commitment for the remaining $200,000 permanent-loan balance. For an additional fee, the mortgage lender may extend his credit to cover the gap. As another alternative, the developer could take out a short-term second mortgage.

The mortgage banker's role, in finding good sources of capital on favorable terms for the developer is crucial.

An increasingly popular method of closing the gap is the standby commitment. This is a viable mortgage-loan commitment issued by mortgage trusts and other lenders specializing in such financing. It is accepted by construction lenders as a basis for advancing loans. The standby commitment differs from the normal mortgage loan commitment in these noteworthy respects:

■ Neither the lender nor the borrower intends to consummate the deal. Both recognize the borrower's intention to seek a more favorable mortgage-loan commitment around project completion time. Some lending specialists dealing in standby commitments would be caught short of funds if more than a small fraction of their clients actually took the long-term loan.

■ The interest rate is set purposely high for a short term, to discourage the borrower from exercising his privilege.

■ Prepayment penalties are waived, allowing the borrower to repay the loan at any time without penalty.

Sophisticated developers sometimes use the standby commitment as a short-term financing technique that allows them to build their projects and then, with a demonstrated income-producing project, to seek a mortgage loan on more favorable terms.

Secondary mortgages of various kinds are available for different developers' needs. Since the first-mortgage lender's claim takes precedence over a second mortgagee's, secondary mortgages carry interest rates ranging from 30 to 80 percent higher than first-mortgage rates. Loan term for a second mortgage is likely to be less than five years.

Equity-sharing financing techniques include the following basic themes:
■ Joint ventures ■ Leaseback arrangements ■ Sale-buyback agreements
■ Development loans ■ Contingent interest

The joint venture is a popular development financing technique. Normally, the institutional investor furnishes all, or most, of the equity investment while his developer partner contributes his entrepreneurial skill. Under such arrangements, the investor resembles the preferred stockholder, and the developer resembles the common stockholder in a private corporation. When the investor provides 100 percent of the cash investment, he usually gets back all his money, plus an appropriate return, before the developer gets a large share of the cash flow.

Under this type of joint-venture deal, the institutional lender should recoup his investment in six to eight years. Thereafter the proceeds are split by contractual formula, possibly 50-50. The developer might arrange to share a substantial part of the depreciation tax shelter. He might also profit from the land sale to the joint venture and from fees for related services.

The sale-leaseback is another popular equity-financing vehicle, in which the developer sells all or part of the project to the investor and then leases the portion he has sold. Among the benefits realized by the developer from the various sale-leaseback transactions are:

■ Improved financial leverage (through a higher financing-equity ratio) ■ Immediate realization of land-sale profits ■ Probable reduction in carrying charges (through elimination of all or part of the amortization cost included in a mortgage debt service payment).

To the owner-investor, the sale-leaseback offers complementary benefits. Under most sale-leaseback arrangements, the owner gains a share in project profits. The owner also benefits from appreciation of land value when the lease expires (see comparison of financing techniques at the end of the chapter).

In the simplest type of sale-leaseback, the developer sells both land and buildings to the investor and then leases them back on a long-term lease— say, 40 years plus two 10-year renewal options. Acting as property manager, the developer subleases to the project's tenants.

Since this type of project in effect features 100-percent financing, thereby heightening the investor's risk, it is normally restricted to well-established developers with good track records. Major terms to be negotiated between developer and lender are lease renewal options, repurchase options, and percentages in rental (or rental increase) split. In some instances, a sale-leaseback is prearranged before construction starts, with the sale made contingent on a given level of rentals—say, 80 percent.

In the land-sale-leaseback, the most popular version of the sale-leaseback technique, the developer sells the land to the investor, but retains ownership of buildings and other improvements. The land-sale-leaseback entails a subtle compromise between the simple sale-leaseback and a conventionally financed project. By selling the land, the developer increases his financial leverage, since he effectively achieves 100-percent financing on the land value, along with the usual 75-percent financing on the improvements. Suppose the project is appraised at $3 million, with land at $1 million and building at $2 million. With a normal 75-percent mortgage— $2.25 million—and a project cost of $2.75 million, the developer's cash investment would be $500,000. If, however, he negotiates a sale-leaseback on the land, he can then negotiate a 75-percent mortgage (known as a leasehold mortgage) on the $2-million building. This cuts his cash investment in half, to $250,000. (He gets a $1.5 million leasehold mortgage loan plus $1 million for the land, requiring a $250,000 cash investment to equal the total $2.75-million project cost.) Moreover, since land is not depreciable, the developer retains the entire project depreciation. For a $2-million apartment building with a 40-year estimated life, a developer in the 50-percent income-tax bracket would gain an additional $25,000 clear after taxes in the first year. (Figured by the double-declining balance method, the first year's depreciation would be 2 x $2,000,000/40=$50,000, half of which would go for federal income tax.)

In the normal land-sale-leaseback, the institutional investor demands an equity share, or "kicker," which usually means one, or more, of the following: a 1- to 3-percent share of the project's effective *gross* income; 10- to 20-percent share of increases in gross income over a predetermined base (usually the projected gross income minus any future tax increase); a percentage of defined income (gross income minus debt-service and operating expenses).

An institutional investor generally prefers to share in *gross* rather than *net* income. Gross income is more easily defined and thus forms a less controversial base for equity sharing.

A land-sale-leaseback transaction is normally executed simultaneously with a leasehold mortgage loan.

High interest rates depress a development project's investment prospects by reducing the mortgage loan amount for which the project qualifies and by raising the owner's operating expenses by a significant amount.

The sale-buyback, also known as an installment sale contract, offers the developer a unique combination of advantages: maximum tax shelter *and* maximum financial leverage. The investor's advantages are more limited, thus making the sale-buyback a less popular financing vehicle than the sale-leaseback. But even for the investor, there are some unique advantages in a sale-buyback.

Like the basic sale-leaseback, the sale-buyback starts with the sale of the property at a price set at 100 percent of the total audited land and construction cost, which usually falls between 80 and 90 percent of appraised market value. (The low selling price gives the construction profit to the developer, a universal practice in development financing.) But then, instead of leasing the project to the developer, as in a sale-leaseback, the investor sells the project on a long-term installment sale contract, which offers the investor a higher yield than he would receive in a typical mortgage loan repayment.

A development loan is a mechanically simple financial technique, restricted to established developers, for acquiring land and installing such basic utilities as roads, water-supply systems, sewers, etc. This short-term loan precedes the construction loan. It carries a high interest rate to compensate for relatively high risk. It usually carries a provision for the investor's equity participation—e.g., a percentage of gross income, an actual equity interest, or possibly a share in the builder's profits.

Contingent interest is an equity-sharing technique normally appended to a mortgage loan at fixed interest rate. In a mortgage-loan contract calling for contingent interest, the developer might pay 10 to 20 percent of annual gross project income exceeding an agreed base amount (usually the gross income projected in the developer's pro forma). Some state usury laws may limit contingent interest payments.

Ad-hoc negotiation of unforeseen dilemmas sometimes turns debt-financing into equity-financing arrangements extracted from reluctant developers who find themselves in a financial hole when the economic ground gives way. A recently completed midwestern office building failed to meet the minimum occupancy requirement to qualify for its committed mortgage loan. On completion of this project, the owners were apparently confronted with two dismal alternatives:

■ Raise a large amount of additional equity capital to keep the project going until occupancy rose to the level required to free the mortgage loan.

■ Yield ownership to the lender.

A compromise solution, proposed by the lender, averted these extremes. In return for extension of the interim loan period beyond its contracted expiration date, the lender took a 15-percent equity share of *gross* project income. The office building soon proved successful, with a 95-percent occupancy. But the lagging start tore a permanent, unrecoverable chunk out of the owners' equity.

Speed in securing a mortgage commitment is vital in times of rapidly escalating construction costs—a good mortgage banker should be able to do it in 45 days.

As the preceding discussion should have amply demonstrated, the architect usually needs a good mortgage banker to guide him through the financing jungle, with its bewildering and proliferating techniques. In addition to the mechanical tasks for arranging and servicing loans, the mortgage banker provides many ancillary services. Before obtaining the mortgage commitment, he reviews project economics and market area. After completing the mortgage negotiations, he checks on details concerning leasing, construction and other aspects of the project development that might affect financing. A good mortgage banker, or someone else fully capable of playing his role, is an indispensable member of the development team.

COMPARISON OF FINANCING TECHNIQUES

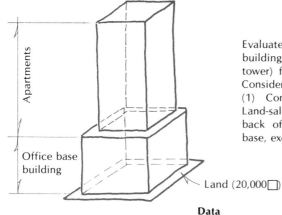

Apartments

Office base
building

Land (20,000☐)

Evaluate the sketched multi-use building (office base plus apartment tower) for *financial leverage only.* Consider three financing techniques: (1) Conventional mortgage; (2) Land-sale-leaseback; (3) Sale-lease-back of land *and* office-building base, exclusive of apartment tower.

Data

Land area: 20,000 sq ft @ $100	=$ 2 million
Office bldg: 200,000 sq ft @ $30	=$ 6 million
Apartment tower: 200,000 sq ft @ $20	=$ 4 million
Total	$12 million

Estimated Income

Office bldg: 200,000 sq ft @ $7	=$1,400,000
Apartments: 700 rooms @ $1,000	= 700,000
	$2,100,000

Estimated expenses

Real estate taxes: $400,000
Office operation: 300,000
Apartment operation: 100,000   800,000

Net income   =$1,300,000

(1) CONVENTIONAL MORTGAGE

Appraised value, capitalized @ 9.5% $=\dfrac{\$1,300,00}{.095}$ =$13,600,000

First mortgage @ 75%   =0.75 x $13,600,000 = 10,200,000

Developer's cash investment   = $12,000,000
                                                      −10,200,000
                                                      $ 1,800,000

Permanent loan: 8.75% interest, 30 yrs, 8 mos; 9.5% Constant

Debt service =.095 x $10,200,000   = $969,000
Cash flow   =$1,300,000−$969,000   = $331,000

% Yield $= \dfrac{\$331,000}{\$1,800,000}$
=18%

(2) SALE-LEASEBACK, LAND ONLY

Land is sold for $2 million (development) cost

Ground rent $=8\%$ x $2,000,000 $=$160,000

Net leasehold income $=$1,300,000—$160,000
$\qquad\qquad\qquad\qquad$ $=$1,140,000

Appraised leasehold value, capitalized @ 9.5% $=\dfrac{\$1,140,000}{.095}$ $\quad=$ $12,000,000

Leasehold mortgage $=75\%$ x $12,000,000 $=$ $\qquad\qquad\qquad\qquad\qquad\qquad\qquad$ 9,000,000

$\qquad\qquad\qquad\qquad\qquad\qquad\qquad$ Land sale $=\quad$ +2,000,000

$\qquad\qquad\qquad\qquad\qquad\qquad\qquad$ Total financing $=$ $11,000,000

Developer's cash investment $=$12,000,000—$11,000,000
$\qquad\qquad\qquad\qquad\qquad\qquad$ $=$ 1,000,000

Permanent loan: 8.75% interest; 30 yrs, 8 mos; 9.5% Constant

Mortgage debt service $=.095$ x $9,000,000 $=$ $\qquad\qquad\qquad\qquad$ $ 855,000

$\qquad\qquad\qquad\qquad\qquad\qquad$ Ground rent $=\qquad$ 160,000

$\qquad\qquad\qquad\qquad\qquad\qquad\qquad\qquad\qquad$ $1,015,000

Cash flow $=$1,300,000—$1,015,000 $=$285,000

$\qquad\qquad\qquad$ % Yield $=\dfrac{\$285,000}{\$1,000,000}$

$\qquad\qquad\qquad\qquad$ $=28.5\%$

(3) SALE-LEASEBACK OF LAND AND OFFICE BUILDING

Land and office building are sold for $8,000,000 and leased back at an annual rent of 9.375%. This transaction is complicated by air rights, which requires some tricky bookkeeping. The developer continues to own the apartment building, and he collects rents from both office and apartment buildings. A $100,000 annual charge for air rights, however, must be charged against the apartment building in calculating its capitalized value. The $100,000 air-rights charge is credited to the office base. Thus its only practical effect is to reduce the apartment mortgage loan to a more conservative figure, to protect the lender's interest.

Base office building		Apartment		

Income $=$1,400,000$+$$100,000 (air rights)   Income $=$ $700,000
$\quad$ $=$$1,500,000

$\qquad$ Expenses

Expenses $\quad$ Air rights $=$ 100,000
$\quad$ Real estate taxes $\quad$ $=$$300,000 $\quad$ Real estate tax $=$ 100,000
$\quad$ Operating expenses $\quad$ $=$ $\underline{\ 300,000}$ $\quad$ Operating $=$ $\underline{\ 100,000}$
$\qquad\qquad\qquad$ $600,000 $\qquad\qquad\qquad$ $300,000

Net Profit $\quad$ $=$$1,500,000—$600,000   Net Income $=$ '$400,000
$\qquad$ $=$$900,000

$\qquad\qquad\qquad\qquad\qquad\qquad\qquad$ Capitalized value $=$ 400,000
Annual Rent $=$.09375 x $8,000,000 $\qquad\qquad\qquad\qquad$ $\underline{\quad .095}$
$\qquad$ $=$$750,000 $\qquad\qquad\qquad$ $=$$4,200,000

Cash Flow $\quad$ $=$$900,000—$750,000   Leasehold loan @ 75% $=$$3,150,000
$\qquad$ $=$$150,000

$\qquad\qquad\qquad\qquad\qquad\qquad\qquad$ Debt service @ 9.5%C $=$ $300,000

$\qquad\qquad\qquad\qquad\qquad\qquad\qquad$ Cash Flow $\qquad$ $=$$400,000—$300,000
$\qquad\qquad\qquad\qquad\qquad\qquad\qquad\qquad$ $=$$100,000

$\qquad$ Total financing $=$$8,000,000 (from land$+$office bldg. sale)
$\qquad\qquad\qquad\qquad$ $\underline{+3,150,000}$ (apt. tower loan)
$\qquad\qquad\qquad\qquad$ $11,150,000
$\qquad$ Developer's cash investment $=$$12,000,000$-$$11,150,000
$\qquad\qquad\qquad\qquad$ $=$$850,000
$\qquad\qquad$ Total cash flow $=$$150,000$+$$100,000
$\qquad\qquad\qquad\qquad$ $=$$250,000
$\qquad\qquad\qquad$ % Yield $=$$\dfrac{$250,000}{$850,000}$
$\qquad\qquad\qquad\qquad$ $=$29%

From Sonnenblick-Goldman

VI

THE ARCHITECT'S ROLE IN THE DEVELOPMENT PROCESS

"The same talent that creates projects can also initiate projects."
— Ronald S. Senseman, FAIA

Harry A. Golemon, AIA, has noted a striking paradox in the profession's structure of services. As part of his basic service contractual obligations, an architect must submit statements of probable construction cost at each stage of the project. Yet the program and financial feasibility studies are often made and approved before the architect is commissioned. If the financial feasibility study contains an inaccurate construction cost estimate, the program and the budget may pose an impossible combination of goals for the architect. The question naturally arises: How can an architect be held responsible for controlling costs if the program requirements cannot be achieved under the budget?

The answer is that the architect should participate actively in the decision stage, when he can contribute his knowledge of construction costs to the market analyst. Among the decision-stage steps discussed in Chapter II, site selection, programming, construction cost projection and schematic design fall within the normal scope of architectural work. Qualified architects also may do economic design and prepare the financing package for submission to prospective lenders. As the developer-entrepreneur, the architect may assemble the development team and acquire the land.

A North Kansas City, Mo., medical office building project designed by Herbert E. Duncan Architects, Inc., illustrates the advantages of the architect's participation in the decision stage. In addition to providing comprehensive professional services, the architect also served as a leader of the development team and consultant on all major project decisions—including site selection, zoning and economic design. A contractor was selected as soon as the preliminary drawings were completed. He joined the development team, which, in addition to the architect, included five doctor owners, an attorney, a leasing agent, a management agent and a mortgage banker.

Architectural programming for this medical building required careful analysis of physician needs, financial resources and the ability of the medical staff of an adjacent hospital to provide occupancy. The architect also had to coordinate his work with that of another architect who was commissioned to design a proposed $9-million addition to the hospital. Rezoning for both buildings was concurrent and unusually difficult due to city ownership of the hospital. Negotiating the necessary changes took many meetings attended by the two architects, neighborhood representatives and city officials.

Throughout the preliminary design period, the Duncan firm worked closely with the contractor on cost control. From detailed preliminary drawings, completed two months after the owners' approval of the initial design, the contractor submitted a guaranteed maximum cost bid of $1.7 million. This figure satisfied the investment analysis and the mortgage banker. Loan papers were signed in June, and construction started on a fast-track schedule on July 1, 1971. In January, 1972, the project was rolling toward its scheduled completion date of July 1, 1972, with a considerable cost saving anticipated.

As a means of assuring design control, however, nothing succeeds so well as sole ownership by the architect. Such a phenomenon generally is limited to small projects. The following case study shows how a principal in a small firm can combine an architectural with the entrepreneurial role.

Architect Harry Wenning, AIA, of Hastings-on-Hudson, N. Y., spotted an opportunity to build a small office building on the outskirts of his suburban town's small business district. During an association with a home builder for whom he was designing houses, Wenning noted an interesting property for sale—an old Victorian house located on a lot transected by a zoning line dividing residential from commercial development. Most of the house, at the front of the lot, was in the residential zone, but part of it, at the rear, lay in a 100-foot-square portion in the commercial zone.

Seeing this opportunity for building a small office building, Wenning and his contractor partner bought the property for $37,000, lopped off the portion of the house in the commercial zone, and sold the remaining residential property for $27,000. This transaction left them with a 10,000-square-foot commercial site acquired for a base price of $10,000 plus legal fees after the building was completed. Wenning bought out his contractor partner, who needed cash for a housing project.

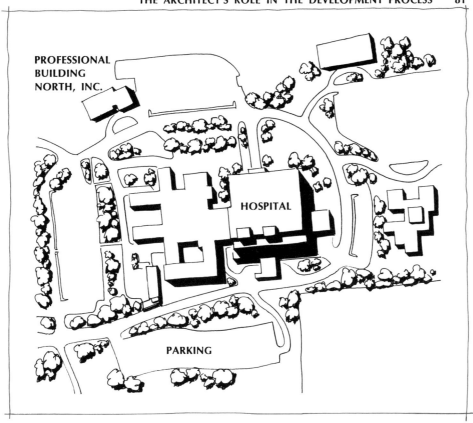

The team for development of the Professional Building North, Inc., included Herbert E. Duncan Architects, Inc., five doctors, an attorney, a leasing agent, a management agent and the contractor. The building is adjacent to the North Kansas City Memorial Hospital.

Wenning needed no formal market or feasibility studies. Knowing of the shortage of office space in Hastings-on-Hudson, he built on speculation. (One sure tenant was his own architectural firm, Wenning & D'Angelo, which needed 1000 square feet.) The development team, in addition to the architect-owner, consisted merely of an attorney and a tax accountant. Wenning himself arranged for financing through a local bank, a contact he had established through earlier experience in house construction. He sub-contracted construction of a two-story office building with 5500 square feet of net office space designed for maximum flexibility in layout, with 40-foot clear spans designed for 75-psf floor liveload; eight independent HVAC zones, each served by a packaged rooftop or ground-level unit, and partitions omitted until the space was leased.

Wenning built on speculation because he felt that the almost immediate availability of office space already built would attract tenants.

"Office lessees in this area have been turned off by construction signs proclaiming x square feet of office space promised by April 1 and then seeing the foundations barely excavated by the date they were scheduled to move. I was so sure of my market that I just went ahead and built," said Wenning.

His faith was well founded. Advertisements in the New York Times procured three lessees: a dentist who took 700 square feet and two medical doctors' groups who took a total of 2800 square feet. A realtor rented the remaining 1000-square-foot space to a group of university program consultants.

On larger, more complex projects in which the architect has little hope of attaining total ownership, his role inevitably becomes more complex and relatively more limited. He nonetheless can achieve the design control that is not assured by a conventional architectural commission.

The neophyte in development work is well advised to use all available technical tools, experience, and judgment. Minimize the role of hunches, intuition, sixth senses, and blind guess.

A hospital project designed by Golemon & Rolfe, of Houston, illustrates the range of development roles open to architects with the required skills. As joint venturer with three doctors who own the majority interest, G & R is taking the developer's role in a 192-bed psychiatric hospital. In addition to standard architectural programming and design services, the firm is acting as development manager. The architects negotiated a lease with a hospital corporation, prepared the financing package, obtained the permanent financing through a mortgage banker and projected long-range investment goals for the joint-venture owners. As construction manager, G & R negotiated a $4.5-million stipulated-sum contract with a general contractor.

Turnkey public housing is a fertile field for architects-developers. Because of the superior design achieved when architects have strong design control, the Department of Housing and Urban Development encourages its district offices to favor architects' involvement in these projects. The following hypothetical case study is typical of this type of development opportunity.

Architect Gerald Maceroy Associates, an eight-man firm in Excelsior, O., initiated a $1.2-million turnkey public housing project with the Excelsior Housing Authority. Maceroy first approached a realtor who owned a suitable four-acre site. A contractor was recruited as a third member of a joint venture to build the project. Preliminary study already had indicated the site's proximity to schools, public transportation, shopping facilities and general suitability for housing. The architect prepared schematic drawings and specifications and the realtor made tentative arrangements for a construction loan. A proposal, based on preliminary drawings and preliminary cost estimates, was submitted to the local housing authority in competition with other design-build proposals.

Turnkey public housing is a fertile field for architect-developers.

After the proposal was accepted, the architect, as the joint venture's representative, wrote the letter of intent and signed the contract of sale, under which the local housing authority agreed to purchase the completed project.

Leadership, as well as ownership, is an important issue concerning the architect's role on a development team. Depending on his firm's experience and qualifications, the architect may or may not be the team leader. The AIA has established guidelines for interprofessional collaboration with other professionals in environmental design. In Professional Collaboration in Environmental Design, a guide published by the AIA and members of the Interprofessional Council on Environmental Design, there appears the following statement:

". . . Ordinarily the client's interests are best served in the research, analysis and design of a project when the client has a single contract with a prime professional who is responsible for direction of the work and for providing through collaboration the specialized services that may be needed. This makes available to the client all the advantages of specialization and at the same time centralizes responsibility."

CONTROLLING DESIGN AND BUILDING COSTS   The one imperative for an architect is to retain control of architectural design. Regardless how many other tasks he assumes, if the architect surrenders control of the design he is defaulting on his professional obligations. Design, after all, remains the heart of architectural practice. Market analysis, construction management, and other non-design services all are ancillary services to be performed in addition to, but not in place of, basic design. The architect may, of course, serve as a consultant for such services to another architect rendering design services. He also may render consulting services as a member of a professional environmental design team, headed possibly by a professional planner or an engineer. The architect should be alert to the possibility of another development team member from outside the design professions usurping his design role.

As indicated earlier, cost control, no less than design control, requires the architect's involvement in the decision stage. This necessity was illustrated by Bernard J. Grad, FAIA, in the book, Creative Control of Building Costs. A large university client presented Grad's firm with an unrealistic program and budget for a classroom and science building. The client's program called for 44,000 square feet of net usable space with an allowance for only 25 percent, or 11,000 square feet, for nonassignable areas—circulation and public toilets. From past experience, the Grad firm knew that such a building required about 19,000 square feet of nonassignable area. After reviewing space needs with the affected department heads, the Grad firm, with the client's consent, altered the program. The minimum net usable area was reduced to 36,000 square feet; total area remained at the original 55,000 square feet, and the project came in about 1 percent below budget.

Happy endings should not, however, obscure the point that it often is wasteful and inefficient for a client to exclude the architect from the decision stage. A good program cannot be written by anyone unfamiliar with space requirements, construction costs, and other aspects of the building process familiar to the architect. A private developer inevitably would suffer more than a university from the kind of programming error encountered by the Grad firm. If a feasibility study incorporating an impractically optimistic estimate of net usable floor space were the basis for preliminary financing negotiations before the architect began his design, the lost efforts and delay in correcting the feasibility study could prove costly.

Design quality, as well as cost control, depend on early involvement by the architect and his consultants. Early involvement by the mechanical and structural engineers elicits advanced design concepts when they can be best investigated and exploited. Efficient packing of the structural framing, lighting-ceiling troffers, and airconditioning ducts into the ceiling space is one of several constantly recurring design problems. Solving this problem may require considerable consultation among the architect, structural engineer and mechanical engineer. The earlier they solve it, the more opportunity for manufacturers to produce suitable building components or subsystems and to satisfy required delivery schedules. If such basic design decisions are not made early in the development process, the owner may have to accept less efficient, more expensive subsystems merely because they can be designed, fabricated and erected more quickly than a more technically advanced and more economical design solution. The ubiquitous pressure to resort to second-rate, but readily available, expedients instead of a more deliberate pursuit of design economy and quality is built into the traditional building process. The widespread practice of excluding the architect and his technical consultants from the decision stage contributes to the building industry's technological inertia.

As Dudley Hunt, Jr., FAIA, explains in Creative Control of Building Costs, cost control requires a continual monitoring starting long before the preparation of design drawings. It has become an indispensable part of the design process itself. When the low general-contract bid is far above the budget, it can damage the creative design process beyond repair. Desperate wholesale substitutions and deletions sometimes are required to save the project, and they inevitably loosen the architect's grip on the design. The hasty revision may result in a pallid vestige of a healthy original. It is far better for the architect to work within the budget throughout the design process than to suddenly confront need for drastic revisions.

Tight design and cost control require, of course, progressively refined cost estimates. They should start with the crude per-square-foot cost projections during schematic design and progress through more accurate in-place, or unit-of-enclosure methods, into the quantity takeoff and pricing estimate for design drawings and specifications. With ample time to incorporate it into total design, an architect may turn a budget limitation into an imaginative, economical design. But if he is faced with a suddenly imposed deadline to revise his design, or confronted early in the design stage with an unrealistic program, he will have less time and opportunity to respond creatively to the budgetary challenge.

ACCELERATING THE DEVELOPMENT PROCESS The architect plays a key role in the fight against the clock and inflationary construction costs that have been climbing at a rapid rate since the 1960s. If he is retained as professional construction manager, the architect is responsible for compressing the construction schedule to hasten the completion date. Even restricted to his design role, the architect must know fast-track and other techniques for telescoping the development process.

The combination of staged bidding with systems building is sometimes the best way to accelerate the development process to the tempo required in today's market. This combination can cut as much as 45 percent from the time required for the development process under conventional lump-sum general contract bidding. Even on moderate-sized—$2 million to $5 million—projects, accelerated scheduling can cut six to eight months from the delivery time. Prebidding of subsystems before general contract awards produces the major time savings. A telescoped decision-design-delivery schedule then enhances the time savings already gained through the systems approach.

A large health science complex at the Stonybrook (Long Island) campus of the State University of New York demonstrates the tremendous time and cost-saving potential of the accelerated team approach. In 1969, under traditional lump-sum general contract bidding, this project came in more than 50 percent over the budget. After rejecting this low bid, the State University Construction Fund called in Smith, Hinchman & Grylls Associates, Inc., as architects-engineers and construction manager to redesign and revamp the entire development process. SH & G applied its version of accelerated scheduling, Unified Team Action Program (UTAP), breaking the project into 21 separate contracts or bid packages. By telescoping the entire development process on a crash program, SH & G produced the complex within nine-and-a-half months, at a budget-beating $12.2 million. Landscaping was completed later (see Stonybrook schedule). This completion date was about 30 months earlier than the estimated completion date for the same project handled under a conventional lump-sum general contract. It pared an estimated $5 million in construction escalation costs, projected at an annual 1-percent monthly rise. And it saved the university an estimated additional $3 million in rental costs that would have occurred under a 30-month delay in occupying the complex.

On staged contracts with decision, design and delivery stages often overlapping, responsibility for construction delays becomes extremely difficult to pinpoint.

Systems building works well with accelerated scheduling for two reasons. The flexibility afforded by relocatable partitions, multi-zone airconditioning, versatile lighting-ceiling subsystems, etc., lightens the pressure of architectural design, allowing late decisions on precise room layout, sizes and other design details. Use of standardized components allows prebidding very early in the development process—sometimes within two months of the architect's commissioning date. (see fast-track systems schedule.)

In addition to accelerating the development process, prebidding of subsystems offers another major advantage in cost control. Assume 50 percent of the estimated project cost is in prebid systems components. These firm prices, set early in the development process, drastically reduce the chances for the remaining contracts to force the construction cost above the budget. What makes the systems prebidding feasible before completion of design is the standard pricing principle for standard components. Since the subsystems are standard—mass-produced components with standard unit prices—it is unnecessary to know the precise quantity and cooling capacity of airconditioning units, the length of ducts, etc., for the prebidding. Approximate quantities can be assumed and prebid on a unit-price basis, with adjustments made after the determination of exact quantities in the final design. The designer gets cost control combined with design flexibility.

In another version of accelerated scheduling more suitable for general contract work, the compressed-accelerated method is used for early bidding and consequent telescoping of the development process. Through use of this method, a $3-million, 160-bed hospital in Baytown, Tex., was designed and built in 16 months, at a time saving of six to eight months over estimated time required for the conventional development process, according to architects Golemon & Rolfe.

Unlike the fast-track system technique, the compressed-accelerated approach features simultaneous bidding on all subsystems before any construction starts. The purpose in simultaneous bidding is to retain one of the advantages in a lump-sum general contract—foreknowledge of the construc-

DESIGN & CONSTRUCTION SCHEDULE FOR THE HEALTH SCIENCES COMPLEX AT STONY BROOK

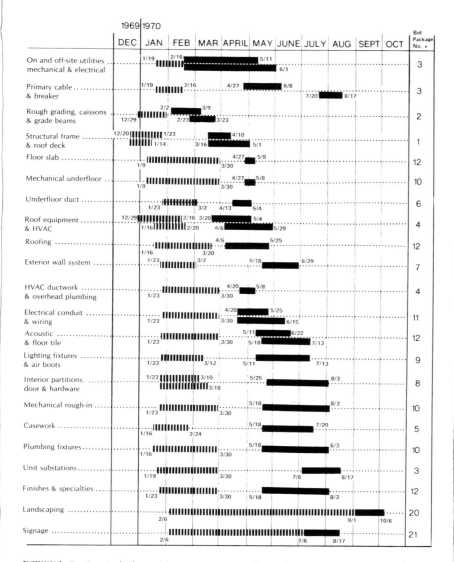

IIIIIIIII Design to bid award   ▬▬▬ Delivery & construction (start to finish)

*Bid package numbers merely indicate the order in which bids were prepared and let. In some cases (unit substations, for example), a component that is installed very late in the job was bid early (No. 3) because of the lead time needed to manufacture and assemble this item.

From *AIA Journal* 7/70

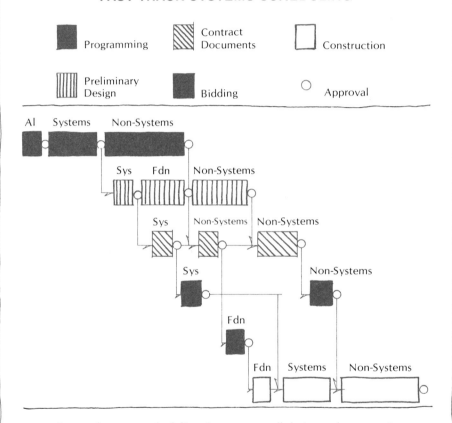

FAST-TRACK-SYSTEMS SCHEDULING

Fast-track-systems scheduling shortens over all design and construction time by over-lapping activities that normally would not be done until the one ahead is completed.

From Educational Facilities Laboratories

tion cost before construction starts. The inevitable price of this advantage is some slight time loss from the potential time savings achieved by staging construction bids.

In the compressed-accelerated method, the bidding process shifts forward into the design development stage. Bid documents comprise complete specifications plus a set of expanded preliminary drawings containing key details necessary for bidding. Later, during the construction stage, the architect finishes the bid documents, filling in details required for construction, though not previously required for bidding. (See chart for a diagrammatic representation of the compressed-accelerated method technique and Professional Construction Management and Project Administration by William B. Foxhall for a fuller description.)

In addition to the role of professional construction manager with responsibility for cost control and scheduling, the architect may serve in yet another role—as project administrator.

Despite their often indispensable benefits in speeding the development process, the accelerated delivery techniques have one notable drawback: they complicate the pinpointing of responsibility for construction delays. Though it is certainly the slowest, and probably the most costly method of constructing buildings, the traditional development process does establish clear lines of responsibility during the construction period. In the various accelerated schedule techniques, the architect becomes much more vulnerable. A stipulated-sum general contract, in which bids are taken on complete drawings and specifications, can place sole responsibility for meeting the construction completion deadline on the general contractor. It can include a simple liquidated-damages clause providing for the owner's reimbursement for losses caused by construction delays. On staged contracts with decision, design and delivery stages often overlapping, responsibility for construction delays becomes extremely difficult to pinpoint.

When the architect serves as professional construction manager, his responsibility inevitably increases over his responsibility as designer. The construction-manager approach usually entails some form of accelerated scheduling. Accelerated scheduling always demands an earlier go-no-go decision by the owner than the conventional development process. Firm, guaranteed prices can be closely approximated, but not assured, in the accelerated process. As the owner's agent advising him on such decisions, the construction manager is deeply involved in cost control. His construction management contract defines the architect's responsibility.

In addition to the role of professional construction manager, with responsibility for cost control and scheduling, the architect may serve in still another role—as project administrator. In this case, the architect acts as the owner's administrative representative in dealing with a contractor. Large corporations engaged in continual building programs have their own construction departments capable of administering the building work. The project

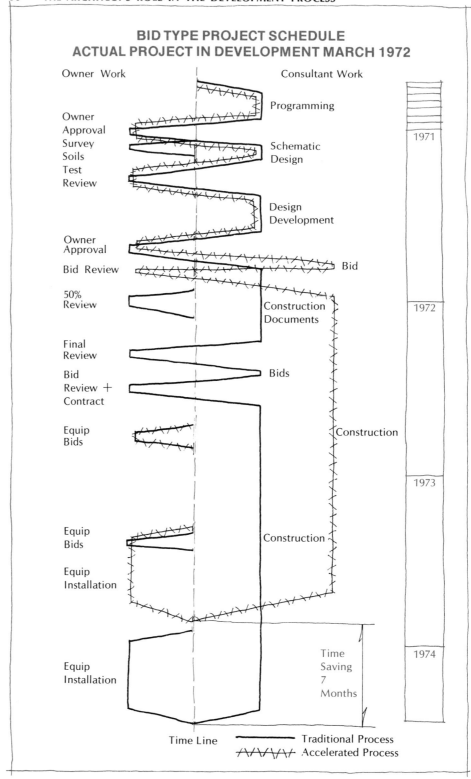

BID TYPE PROJECT SCHEDULE
ACTUAL PROJECT IN DEVELOPMENT MARCH 1972

Owner Work   Consultant Work

                                          Programming

Owner
Approval
Survey   Schematic
Soils   Design
Test
Review

                                          Design
                                          Development

Owner
Approval

Bid Review   Bid

50%
Review   Construction
                                          Documents

Final
Review

Bid
Review +   Bids
Contract

Equip
Bids   Construction

Equip
Bids   Construction

Equip
Installation

Equip   Time
Installation   Saving
                                          7
                                          Months

1971

1972

1973

1974

Time Line   ———————— Traditional Process
                 ⫢⫢⫢⫢⫢ Accelerated Process

administrator's role was created to fill the temporary needs of owners whose construction work is too infrequent to warrant the hiring of permanent personnel.

A FURTHER WORD   Participation in the development of new communities provides a score of opportunities for members of a development team. Provision for federal loan guarantees under the New Communities Acts of 1968 and 1970 has added incentive for establishing new towns through an orderly growth pattern. This sort of development can be achieved privately, publicly, or as a joint public-private venture. Whatever the method, the opportunities for architects are broad.

Building new communities at best is complex—at less than optimum conditions it can be incredibly complicated. Washington, D. C.'s Fort Lincoln New Town is a case in point.

Fort Lincoln, as conceived and initiated by Lyndon B. Johnson in 1967, was to be the model for a national program of surplus land development into model housing for middle- and lower-income families. It began as a good idea—and with perseverance—will fulfill its promise by 1976. But the years between start and finish have been difficult.

At the outset, overlapping governmental authority was evident. In addition to the White House staff, the National Capital Planning Commission, HUD, and the District of Columbia's municipal government, there were the Redevelopment Land Agency and the National Capital Housing Authority. Add to these the new community's neighbors and the consultants—architects, engineers and economic and transportation consultants. All were dedicated to the same project—transforming a 320-acre tract of publicly held land into a complete, new, socially and economically balanced community.

Hampered from its birth by overlapping jurisdictions and a general confusion as to goals, the Fort Lincoln project—future home for 25,000—bogged down in a maze of complicated details which left the design professionals wondering who, indeed, the client really was. The citizens felt they had been given an inadequate voice in the planning, and the governmental authorities tried to please everyone.

The result? The project barely moved during its first four years.

For Fort Lincoln, however, 1972 began as a good—and promising—year. Richard M. Nixon announced renewed support for the program and has announced that his Administration plans to have much of the Fort Lincoln project completed by the 1976 Bicentennial. The dedicated men who began the project in 1967 will benefit from the lessons learned during the initial frustrating planning period.

Another example of development of new communities—in the private area and on a different scale—shows how one A-E firm looked at the problems first and tackled the project second.

In the mid-1960s, Reynolds, Smith and Hills, a broad-based professional

The Oak Forest Apartments near Gainesville opened its initial 240 units of two-, three-, and four-bedroom garden and townhouse apartments in mid-1972. Nine of the principals of Reynolds Smith & Hills formed a joint venture with three other developers to form Bivens Arm Developers. RS & H manages Bivens Arm Developers and also provides operational management for the apartments.

services firm in Jacksonville, Fla., established the RS & H Planning Division that served many clients—large and small—in the land development industry. Early in 1970, as a result of the firm's increased involvement in site selection, market and site analysis and comprehensive planning, the firm took the further step of involvement in new communities development in an ownership position. Nine principals of RS & H entered into a joint venture with a group of landowners in the Gainesville, Fla., area to build a large apartment and condominium complex as a substantial part of a developing area. A Jacksonville financial and development company acted as a catalyst and served as mortgage banker.

Each step in the birth of this new communities project was carefully detailed and evaluated by the development team in successive order. Economic design, coupled with studies of community relationships, involved extensive market analyses and feasibility studies. Program analysis and budget projections were carefully developed before starting schematic studies. The development of schematics was done cautiously with constant testing against the economic design and social and political climate. With the financing package in hand, a permanent financing commitment was secured after an exhaustive search for the most favorable interest rates.

After construction was underway, with initial occupancy scheduled for mid-1972, RS & H assumed the apartment management responsibilities. A study of this development project shows how a knowledgeable design firm can apply its talents to participation in a new communities venture through all its phases: decision, design, delivery and even post-delivery.

Programming and budgeting decisions made in the decision stage without the architect's participation can strip him of effective design control. But decisions made carefully, and with full participation of all members of the development team, can produce successful joint ventures.

BUSINESS ASPECTS

"Ultimately, professional responsibility is ineffective without business responsibility."
— Harry A. Golemon, AIA

As a co-owning equity participant in a development project, an architect faces several major problems avoided by architects providing only professional services. One is the collective problem, shared with his co-owners, of forming the most profitable and legally secure business organization. The architect also confronts the personal problem of how to translate his professional compensation into a reasonable share of the equity. There is also the problem of separating his business role from his professional role.

ORGANIZING FOR DEVELOPMENT An architect can choose one of three basic forms of developer organization: Sole proprietorship, partnership or corporation.

A fourth type of organization, the real estate investment trust (REIT), is primarily a means of diversifying the risks of many projects. A REIT normally is formed by a large land developer, a mortgage lending institution, or mortgage broker.

Operating on a small scale, architects who become sole owners of small projects—office buildings or retail stores, beach houses for sale—may own these properties simply as individuals.

Partnership is the most popular form of development organization. As its chief advantage, a partnership is not a taxable entity like a corporation. Income is taxed to individual partners. They receive all tax benefits—depreciation tax shelter, deductions for interest payments, etc., and partners must pay taxes on all income whether or not it is actually distributed.

The joint venture is a popular form of partnership for architects in development work. A joint venture is an ad-hoc partnership, created for one specific project or series of projects. It is recognized as more limited in scope than either general or limited partnerships, which are formed for more general, continuing enterprises. A typical joint venture might comprise an architect, a contractor, a tax accountant and a real estate broker, associated as co-owners of a shopping center. They might plan to sell the project in 10 or 12 years, when the tax depreciation shelter runs out, and terminate the joint venture. In the meantime, each would own equity shares, similar to a business corporation's common stock. The shares could be sold by one joint venturer to the others, or to an outsider. Thus joint-venture investment offers some liquidity, though obviously less than common stocks traded daily on national or local exchanges.

In the choice between a general or limited partnership, the limited partnership may offer the more useful vehicle for development work.

For a continuing series of development projects, a general or limited partnership may prove more efficient than a series of joint ventures. Both offer greater flexibility than one or more joint ventures, since neither is limited to the specific objectives that restrict joint venture operations.

In the choice between a general or limited partnership, the limited partnership may offer the more useful vehicle for development work. A limited partnership provides for the addition of passive investors—the limited partners—as a source of additional equity capital, a vital need in many development projects. These limited partners are denied the active management role of general partners who, in either a general or limited partnership, share the authority to act for the others in carrying on the partnership's business. Like their rights, the limited partners' financial liability is limited —to their financial investment in the firm. (General partners are each personally liable for the partnership's debts and other legal obligations.) Limited partners are thus satellites added to the general-partner nucleus, contributing only capital to the enterprise, affecting neither the managerial authority nor the responsibility of the active general partners.

Big architectural firms with a continuing program of land development operations usually form independent corporations for the purpose. John Portman & Associates created Portman Properties, Inc.; Smith, Hinchman & Grylls Associates, Inc., has formed the SH&G Development Corporation. Perkins & Will has its own development-corporation subsidiary, and the former Dallas firm, Harrell & Hamilton, under its new name, Omniplan,

Inc., has one of five divisions devoted to real estate development and mortgage brokerage.

Compared with a partnership, a development corporation offers several advantages. Because a corporation is a single entity, existing apart from its owners, the individual owners' legal liability is limited to their equity interest in the corporation, like the limited partners in a limited partnership. Unless it is limited by charter, a corporation can readily diversify investment holdings, thereby reducing stockholders' risks.

The corporation's chief financial disadvantage concerns the double taxation of its income—first when it is received and then when it is distributed to shareholders. Double taxation is averted when accelerated depreciation shelters the income which can be distributed untaxed to shareholders. On such matters, an architect should consult a tax accountant or attorney.

Most developers favor incorporation, but some use a combination of incorporation and limited partnership. To reduce personal liability, a developer may incorporate his continuing business operations—e.g., brokerage, leasing, building and ground maintenance, and possibly construction. To reduce his tax liability, he may choose joint ventures or limited partnerships for building and operating each separate development project.

METHODS OF EQUITY PARTICIPATION   An architect who takes all or part of his fee as an equity share in a development project should follow four basic rules:

■ Insist on fair evaluation of services in relation to other development team members' contributions.

■ Delay the entry of other equity investors as long as possible on projects initiated by the architect, either as sole or part owner.

■ Take an equity position only on an individual basis, or through a separate organization created for development enterprise.

■ Beware of financial risks and tax problems associated with investing the entire professional compensation, or a major portion, in the project.

Though it may seem elementary, the advice for architects to demand fair evaluation of their services stems from some painful lessons learned by architects who pioneered in development work. The temptation, sometimes aggravated by pressure, for the architect to reduce his normal professional compensation is greater in development work in which he shares a piece of the action than in normal architectural work for which he receives a cash fee. Some developers, eager to minimize their cash investment, shop for cheap architectural-engineering services. One large office-building developer reportedly claims that he sometimes can cut his cash investment to zero by cutting the A-E fee by 0.5 percent of the project cost. The architect lured by the potential profits of equity participation into accepting a reduced fee is foolish. Real estate speculation entails risks along with profits. In translating his compensation into equity shares, the architect should demand a fee in line with the fees received by others. Is the lawyer, or the developer, accepting a reduced fee for his equity share? Not likely.

Fair apportionment of equity shares in a joint-venture project requires mathematical precision, not vaguely defined tradeoffs. Suppose, for example, that a joint-venture development team comprises a landowner, a developer-contractor, an architect, and a mortgage banker, all investing their fees or profits in a project requiring $200,000 cash investment. If practicable, a specific dollar value could be placed on each team member's contribution to the project. If the landowner contributes land valued at $80,000, he should get a 40-percent equity share; if the architect's fee is $50,000, he should get a 25-percent share, and so on. The architect should be highly suspicious of vague, unpriced tradeoffs—e.g., a 30-percent equity share for the developer's entrepreneurial services, without a breakdown of evaluated items. In short, the architect should reject a barter system and demand a money system for apportioning equity shares.

The psychiatric hospital mentioned in Chapter VI illustrates how the architect's compensation can be augmented for additional services performed during the decision stage. In addition to its compensation for the typical architectural-engineering services, Golemon & Rolfe received a separate developer's fee (for project administration and construction management), paid in installments to a maximum $50,000 total. This fee is paid as follows: $10,000 in $2000 monthly installments during the decision stage and before mortgage financing is committed; $40,000 balance paid out of the construction loan, as part of the project's direct costs, in monthly installments during the design and construction stages.

G & R also qualified for an additional incentive fee of $50,000 to be paid in cash, a percentage of owners' equity, or a combination of both.

A simpler formula was used for a developer's fee on the previously cited medical building in North Kansas City, Mo., by Herbert E. Duncan Architects, Inc. In addition to the conventional A-E services, the Duncan firm provided consulting services on all development decisions—including site selection, zoning, financial arrangements, and contract negotiations for an additional 0.75 percent of the construction cost.

After he has cleared the fair-apportionment hurdle, the architect confronts another problem that is far less likely to occur in his routine work. When the architect is retained by an ordinary client from whom he receives a fee, he can bill this client an extra charge for additional work beyond that included in a standard agreement for architectural services. But when the architect is also his own client, he may find it necessary to prepare additional promotional material or do other work beyond his basic professional service contract with the development team. As a generalist in the development process, the architect is more likely to perform informal ancillary services than other team members. According to Frank Knoble, vice president of Deeter Richey Sippel Associates, "Very careful accounting is required to insure that the architectural contract is not charged time that should be charged to the client role."

An architect-developer who initiates a project should delay the recruitment of other equity investors as long as possible. This policy has two beneficial consequences:

■ It maximizes the architect's design control during the crucial early part of the decision stage.

■ It maximizes the architect's equity share as sole or co-owner.

The financial advantage of delaying recruitment of new equity shareholders manifests an elementary investment principle: equity shares vary proportionately with risk. When investors come into a real estate deal in the early stages, they naturally tend to demand a larger piece of the action for the same cash investment than when they come in later. If the original investors handle the project's economic design, negotiate the land-purchase option and obtain a mortgage commitment, they can reasonably raise the proportionate value of their original cash investments to reflect the added investment security. If, for example, two original investors each contributed $20,000 at the project's inception, they might evaluate their early entrepreneurial efforts at an additional $10,000 each for performing the aforementioned tasks, thus raising their equity shares to $30,000 each. If two later investors each joined with a $20,000 cash investment, they would each get only 20 percent of the action, compared with 30 percent for each of the original investors. The process resembles the stock market, where unknown, high-risk stocks sell at low prices in the company's early life and then rise after the company has a track record that enhances its reputation for security.

In a development project, a principal in an architectural firm could take his equity partnership on an individual rather than a firm basis. For an architect whose entrepreneurial participation is sporadic rather than regular, his equity sharing may be in his own name. For architects whose development work is a continuing sideline—as Portman, Luckman, etc.—such work can be divorced by the previously discussed separate development corporation.

The reasons for separating the architectural firm from the development firm are to keep the accounting straight, and thus avoid confusion between architectural and entrepreneurial functions, billings, payments, etc., and to preserve the design firm's professional image, shielded from the popular stereotype of the wheeling-dealing promoter-developer.

In pursuing his entrepreneurial role, as opposed to his design role, the architect should refrain from making professional decisions within the architectural firm on matters relating to business decisions of the joint venture.

In assessing the financial risks of investing all, or a major part, of his professional compensation in a development project, the architect should recognize his financial function. On projects in which he exchanges his professional compensation for an equity share, the architect is financing cash flow during what is normally the project's most financially precarious period. Before agreeing to do so, he should investigate rigorously the project's financial prospects, possibly through his own independent investment analysis, to make sure that his prospective return is commensurate with the risks. For advice on this and tax problems associated with this deferral of professional compensation, he should consult a tax accountant or attorney.

ETHICAL CONSIDERATIONS

"Professional status is . . . an implied contract to serve society, above all specific duty to client or employer, in consideration of the privileges and protection society extends to the profession."   —W. E. Wickenden

A major obstacle to architects' participation on development teams is the previously cited fear of violating professional canons of ethics. The AIA headquarters has received hundreds of inquiries revealing widespread misconceptions about imaginary ethical barriers to architects' participation on development teams. These misconceptions flourish among the smaller, more conservative firms. Reinforcing this ethical confusion is a vague esthetic distaste—the architect's belief that topflight professionals do not participate on joint-venture development teams because such work is professionally degrading. The distaste of some architects for the business aspects of building construction is something like the political scientist's distaste for active partisan politics.

   This confusion is understandable. Under the accelerated, post-World-War-II changes in the building industry, the client-agent relationship characteristic of traditional architectural practice has been drastically modified. Institutional clients have replaced personal, single-owner clients, blurring the lines of communication with the architect. Accompanying the increasing size and complexity of construction projects, the proliferation of professional consultants has complicated interprofessional relations, inspiring such publications as Professional Collaboration in Environmental Design by the Interprofessional Council on Environmental Design (ICED). Proliferation of the architect's non-professional competitors poses far greater ethical challenges. Non-professional building consultants — space programmers, space planners, cost consultants, construction consultants, market research firms and large corporations new to construction—generally operate free of professional constraints. So do package builders, the building industry's para-professional response to the owner's demands for accelerated construction schedules and cost control.

THE ARCHITECT'S OBLIGATIONS   Despite the apparent complexity of the ethical problem, however, it can be readily resolved, preserving the architect's ethical relations with his client, with the public, and with his fellow professionals and the building industry. The architect's obligations to the public can be served at least as well in the development role as in the traditional architect's role as the client's agent. As a key participant in the decision stage of the development process, he gains a louder voice ". . . in the effort to improve the human environment," which he is instructed to promote in The AIA Standards of Ethical Practice. None of his four Obligations to the Profession and the Building Industry is affected by his participation on a development team. In fact, the only paragraph so affected, under Obligations to the Client or Employer, is paragraph seven:

"An architect shall not undertake any activity or employment, have any significant financial or other interest, or accept any contribution, if it would reasonably appear that such activity, employment interest or contribution would compromise his professional judgment or prevent him from serving the best interest of his client or employer."

An interpretation of this paragraph, issued in 1971 by the AIA Board of Directors, flashes the green light to architects on development teams. It says:

"This standard provides that the architect must avoid any activity which would put, or which could reasonably be construed to put, his financial interest in competition with that of his client; activities during the construction phase are particularly sensitive to such conflicts.

"He may not engage in building contracting where compensation, direct or indirect, is derived from profit on labor and materials furnished in the building process.

"He may engage in construction management as a professional for professional compensation only.

"As a participating owner of a project, he may perform in any role legally consistent with the position of ownership.

"A real or apparent conflict of interest must be resolved in the best interest of the client.

"The Institute holds that the Standards of Ethical Practice are compromised when a member is employed by individuals or organizations offering to the public architectural services which are in any manner inconsistent with these Standards."

In many situations, ownership is the key to resolving ethical conflict. Obviously, there are no ethical obstacles to an architect's functioning as programmer, designer, market analyst, financial analyst, construction manager, or in any other consulting capacity, either as a co-owner or not. But some joint-venture arrangements that are ethical for an architect-owner are unethical for an architect who is not an owner. If, for example, an architect joint ventures with a contractor and a real estate broker to build an office building for an owner-client, he creates an unresolvable conflict of interest. As a member of such a design-build development team, the architect would profit as a co-venturer from any construction economies effected by the contractor at the expense of building quality. Thus his professional interests as project designer would collide head on with his financial interests as builder-developer. Such an arrangement would constitute a classic violation of his ethics. But if the architect participated as a joint venturer with the same partners in a project leased, not sold, to the same client, then the conflict of interest would disappear. In this latter case, the architect's professional and financial interests are consistent: as co-owner, he has the same interest in quality that he has as designer retained by the joint venture in which he shares ownership.

FOLLOW THE RULES   For guidance in the more complex arrangements characteristic in joint-venture enterprises today, architects should inform the client of any personal financial interest (apart from professional compensation) in the project; and contractually separate the design consultant function from the development team.

Nothing in the concept of professionalism bars an architect from participating in a project development team.

Both rules are designed to resolve conflicts of interest, real and apparent, between the architect as professional and the architect as entrepreneur. The first condition, requiring the architect to notify his client of any financial interest he may have in a project, should allay rational suspicion that the architect may be favoring his financial interest over his professional interest. As part owner of an office building under construction for a lessee, the architect as businessman might be suspected of cutting costs at the expense of building quality, which the architect as professional is pledged to provide. He should inform the lessee about his financial interest, even though the lessee is not his client. By so doing, the architect clears the air and clarifies his dual role. In a sense, this condition transcends the subtleties of professional ethics; it is more a matter of simple candor.

From the technical viewpoint, the second condition, requiring the architect to render his professional services as a separate entity contractually apart from the development team, is perhaps more important. At first glance, such a divorce may appear unnecessary. Under greater reflection, however, its necessity becomes more apparent. Isolating the design function distinguishes the architect's two separate roles, enabling him to function purely as design professional, on one hand, and a co-owning member of the development team, on the other. By officially severing his professional role from his business role, the architect avoids some ambivalent situations that might compromise his professional status if he provided professional services merely as a member of the development team and not as a design professional retained for the service. By commissioning the architect as the design professional, or *one* of the design professionals, the developer preserves the traditional clear-cut client-agent relationship that characterizes the architectural profession.

The need to contractually isolate the architect-developer's professional role from his business role arises when there is a third party involved in the project. Consider the scenario in Chapter II. A banker, a realtor, and an architect form a joint venture to build and lease a research-and-development building to a drug company. Although the architect would design this building to suit the drug company's needs, he would nonetheless render his professional services to the joint venture, not to the drug company. To clearly establish this role, the architect should render these services as design consultant to the joint venture, thus clarifying his status to all parties involved in the project. If he is not formally retained as architect by the development team, but instead renders architectural services as a member of the team and an equity participant in the project, his professional role may appear subordinate to his business role. The AIA Standards of Ethical Practice clearly state that "An architect shall not undertake any activity . . . if it would reasonably *appear* [italics added] that such activity . . . would compromise his professional judgment. . . ."

THE CONFLICT OF INTEREST   This directive to avoid apparent conflicts of interest suggests another warning concerning an architect's entrepreneurial role. If an architect's financial interests in a given class of projects grow so large that he becomes a substantial market force, then his financial interest in that field may subvert, or at least appear to subvert, his professional interests. Suppose, for example, that an architect holding substantial financial interests in a line of nursing homes (or restaurants, or motels) were retained by a business competitor to select a site, do financial feasibility studies, and design a nursing home in the same general area where the architect's projects were located. This situation creates a direct conflict of interest. Can the architect candidly assess the financial feasibility of a competitor's project, knowing that a successful project designed for this competitor may drive his own project out of business?

This question suggests that the architect must stop somewhere this side of total entrepreneurial brinkmanship. He must decide whether architecture or entrepreneurial development is his primary role. If he remains primarily an architect, he should limit his equity participation in large-scale commercial projects to a minority role. Because once he becomes a substantial market force in a class of commercial projects, he confronts a real conflict of interest in advising clients who loom as his competitors. If he holds a substantial interest in a chain of local restaurants, he may be ethically suspect in rendering professional services to other restaurant owners. In that event, perhaps the architect's most prudent course is to eliminate competitors' restaurants from his architectural practice. At the least, he should inform such a client of his financial interest in competing restaurants.

The vital need to separate an architect's professional from his entrepreneurial functions springs directly from the basic concept of professionalism. In seeking clients, architects—along with doctors, lawyers, and engineers—are ethically forbidden to advertise in a self-laudatory manner. This ethically prohibited practice is not merely tolerated but praised in business.

Inform the client of any personal financial interest (apart from professional compensation) in the project. Contractually separate the design consultant function from the development team.

Professionals can't be allowed to operate solely by the rules of the marketplace, for they possess specialized knowledge and skills that are vital to the daily operation of society. We expect the professions themselves to deal with incompetent surgeons, with unscrupulous lawyers, or with unprofessional architects who place their own interest above their clients' interests.

But as the foregoing discussion has attempted to make clear, nothing in the concept of professionalism bars an architect from participating on a development team. By scrupulously following the two conditions designed to resolve both real and apparent conflicts of interest whenever there is a reasonable suspicion of such conflicts, an architect can enjoy an equity-sharing position on a development team, secure in the knowledge that he can fulfill the highest standards of professional ethics.

PROFESSIONAL LIABILITY

"The watchwords of care, caution and competence might well be good slogans for architectural offices to imprint upon their drafting boards."
—George M. White, Jr., FAIA

The liability problems confronting the architect-developer are considerably more complex than the ethical problems, and may pose a greater deterrent to his assuming an equity-sharing role on a development team. This warning should not discourage the architect from participating in development work. These problems can be solved. But the architect should enter the field warily, with his eyes wide open, realizing that his liability is greater than that of architects who work within the traditional scope of professional practice. The need for legal counsel scarcely can be exaggerated.

Associated with potential liability are several other legal problems, concerned chiefly with organizational structure. This choice is based on legal as well as financial considerations. Should the architect form a separate corporation, legally distinct from his architectural firm, for his role as developer? Should he participate on development teams as an individual or as a corporation? Should the development team be a joint-venture partnership or a corporation? The answers to such questions will vary, depending on firm size, state laws, extent of the architect's participation in development work, firm income, and even the architect's personal financial burdens and life-style. For all his legal problems, including professional liability, the essential advice to the architect-developer is to retain an attorney familiar with these problems.

THE LIABILITY PARADOX As co-owner of the project he designs, the architect-developer's chief legal problem concerns the reduction in his professional liability insurance coverage. The need for this type of coverage has grown in recent years. Professionals—notably doctors, accountants, engineers and architects—have become increasingly popular targets for a lawsuit-conscious public. Since its inception in 1958, the premiums for architects' and engineers' professional liability insurance have skyrocketed. Increased by 50 percent in 1969, 20 percent in 1970, and 38 percent in 1971, the basic premium rate multiplied nearly 250 percent in just three years (see liability insurance charts). Yet, despite this rising cost, the vast majority of architects—about 90 percent, according to the largest carrier's estimate—retains this vital coverage. The financial risks resulting from malpractice lawsuits are too great for most architects to assume on their own.

As an example of the problems created by an architect's participation in development work, the architect's relationship with a contractor-partner on a joint-venture team creates potential difficulties for the insurer. Suppose, for example, that the insurer wanted to counter a lawsuit brought against its insured architect by suing the contractor. The architect in a traditional arms-length relationship with a contractor may feel much differently towards such a lawsuit than an architect in partnership with the contractor.

Architects should note, however, that joint-venture turnkey projects, in which they are associated with contractors, currently are uninsurable. In fact, any building project that eliminates the traditional adversary relationship between architect and contractor probably will nullify the architect's professional liability coverage. Integration of design-build functions sometimes makes technical sense, but, by involving the architect in the contractor's errors, it heightens his vulnerability to liability claims. However, solutions to the problem of liability insurance coverage for such turnkey projects are being sought, and the architect is advised to check with his insurance carrier before committing himself contractually.

Far more complicating for an architect-developer's professional liability insurance, however, is his dual role as designer and co-owner. Consider the following case: An architect owns a 30-percent equity interest in Apex Developers, a joint venture comprising the architect, a mortgage banker and a real estate broker. The owner, Apex Developers, decides to sue the architect, who had been retained to provide A-E services, for malpractice in the airconditioning design. Thus the architect would theoretically be suing himself. Since owners constitute the largest groups of claimants against architects, the insurance carriers have, of course, eliminated this bizarre possibility. Incorporated in the typical architects and engineers professional liability insurance policy is the following exclusion:

"This insurance shall not apply . . . to claims made against the insured and claims expenses arising therefrom:

(a) by a business enterprise (or its subrogees or assignees) that is wholly or partly owned, operated or managed by the insured or in which the insured is an officer; or

(b) by any employee, his executor, administrator or next of kin (or his subrogees or assignees) of such business enterprise."

Though the foregoing clause nullifies the architect-developer's coverage for a lawsuit brought by the owner, it does not exclude other kinds of lawsuits. The architect retains his protection against contractors' claims, the second largest class of lawsuits, and also against workmen's claims or property damage claims by others not party to the development team or construction contract. But even in these cases, the architect suffers at least a potential loss in his protection. A notable instance concerns the contractor-to-owner-to-architect triple play. A contractor may, for example, sue an owner for compensation for extra work, and the owner, in turn, may sue the architect as the ultimately responsible party. Under the exclusion, an owner or co-owner architect-developer has no coverage against such claims.

For all his legal problems, including professional liability, the essential advice to the architect-developer is to retain an attorney.

The coverage denied under the current exclusion is partially restored in a new policy underwritten by the largest architect-engineer professional liability insurer, the Continental Casualty Company of Chicago, Ill. Under these new insurance policies, administered by Victor O. Schinnerer & Co., Inc., of Washington, D. C., the architect's protection against an owner's claim is reduced by the architect's proportionate share of ownership: i.e., if the architect holds a 30-percent equity share in a joint-venture project, his liability coverage against owner claims is reduced by 30 percent, and the deductible amount, ranging from $2000 to $100,000, would be subtracted from the remaining 70 percent of the liability coverage. Thus, if a $100,000 liability loss claim were settled between the insurer and the non-architect members of the joint venture, the total payoff might be $65,000 (70 percent of $100,000 minus $5000 deductible) instead of $95,000 ($100,000 minus $5000 deductible). Note, however, that this coverage is limited to claims for remedial work. It does not provide payments for consequential damages, such as rental income loss.

In addition to reduced coverage, the architect-developer runs heightened liability risks as an equity-sharing joint venturer. As a professional designer associated in a joint venture with lawyers, real estate brokers and others who, viewed in a construction industry context, are non-professionals, the architect-developer becomes more of a legal magnet than the architect engaged in conventional architectural practice.

Consider, for example, the following situation: A joint-venture development team comprising an architect (A), an owner-manager rental agent (B), and a contractor (C) designs, builds, and operates a $2-million office building. After the building is occupied, the tenants complain about inadequate airconditioning and threaten to break their leases. As members of the joint venture, both B and C may turn to A, charging him with sole responsibility for an airconditioning failure attributable to poor design. Yet under the concept of "joint negligence," all three may have been at fault—A for inadequate design, B for inadequate maintenance, and C for poor installation. As an independent agent offering professional services only, the architect is less liable to be burdened with the errors of the owner-manager and the contractor than when they are his joint-venture partners. If the cost of rectifying the defective airconditioning system were $100,000, there would be a tremendous difference between sole liability and, say, one-third liability, as determined under the concept of joint negligence.

An architect-developer's professional liability insurance coverage differs in another essential respect from his coverage on conventional projects in which he holds no equity interest. On any project in which he is a joint venturer, the architect must take out a separate policy limited exclusively to that project. This separate joint venture is totally divorced from his continuing basic professional liability policy which covers his conventional work. Architect members of joint-venture design teams comprising several independent professional firms allied on large projects are similarly required to take out a separate professional liability policy.

On any project in which he is a joint venturer, the architect must take out a separate insurance policy limited exclusively to that project.

The insurer has several reasons for isolating liability coverage for joint venturers on design teams as well as development teams. Most important is the possibility that the insured architect may be the only insured member of the team. Regardless of the merits of a case, insureds tend to draw lawsuits more than non-insureds. Joint ventures thus compound the insurer's risks: the insured professional retains total responsibility for his own acts, plus whatever may rub off from joint venturers who allow their liability coverage to lapse or otherwise fail to insure themselves.

Another reason for separating joint-venture projects is to isolate their more complex liability problems from the simpler problems arising in the course of conventional work. Anything that can happen on a conventional project can also happen on a joint-venture project. But the converse is not true. The often tangled lines of responsibility in a joint-venture project creates a whole class of problems that cannot occur on simpler projects with fewer professional or entrepreneurial associations.

Isolation of joint-venture liability contracts offers several advantages to the insured architect as well as the insurer. With each joint venture, the architect starts with a clean slate. His liability record exists apart from his overall liability track record. A poor track record can, of course, raise his insurance premium. As a second advantage, a joint-venture liability insurance policy continues the coverage through a post-construction "discovery period," designed to protect the architect for a time interval exceeding the statute of limitations in the state where a claim may be made, the state in which the project is located, in the state where the architect's home office is located or any other state where jurisdiction over the architect may be obtained. These statute-of-limitations periods range from four to 20 years in the various states that have enacted them. Two discovery periods—one for six, the other for 12 years—have been provided to continue joint-venture coverage beyond the project's completion date without requiring additional annual premiums. At the end of the discovery period, coverage can be provided by endorsement to the architect's master policy but limited to his percentage of ownership in the joint venture.

Here in this area of professional liability and other legal problems, the architect is least advised to attempt a do-it-yourself approach. He should seek legal counsel in all aspects of changing liability coverage and other legal matters associated with development work.

Architects and Engineers

Professional Liability Insurance

AVERAGE PREMIUM RATE LEVEL INCREASES

Base Year—1960

Year	Premium Rate Level
1960	100.0%
1961	100.0%
1962	130.5%
1963	134.0%
1964	142.8%
1965	142.8%
1966	142.8%
1967	157.8%
1968	236.7%
1969	284.0%
1970	391.9%

The above table illustrates the cumulative percentage rate increases for the period shown.

Architects and Engineers

Professional Liability Insurance

AVERAGE CLAIM VALUE

Base Year—1960

Year	Average Claim Value
1960	100.0%
1961	69.9%
1962	106.1%
1963	89.9%
1964	87.4%
1965	90.0%
1966	121.2%
1967	119.9%
1968	125.6%
1969	146.0%
1970	155.8%

The average claim value is the average amount of claim payments and claim expenses of each claim occurring in each year of the program.

The results include all insured firms (approximately 10,000 firms as of 1971) and are expressed as percentages of the base year of 1960 to reflect the upward trend of the severity of the average claim.

Architects and Engineers

Professional Liability Insurance

FREQUENCY OF CLAIMS PER 100 FIRMS

Calendar Accident Year	Frequency Per 100 Firms
1960	12.7
1961	13.1
1962	14.6
1963	13.2
1964	15.4
1965	17.6
1966	17.7
1967	18.8
1968	18.7
1969	20.0
1970	20.6

Frequency of claims per 100 firms is the average number of claims incurred by each 100 firms.

Architects and Engineers

Professional Liability Insurance

ULTIMATE INCURRED LOSS

Base Year—1960

Year	Ultimate Incurred Loss
1960	100.0%
1961	92.3%
1962	222.5%
1963	173.3%
1964	192.0%
1965	227.2%
1966	318.5%
1967	356.0%
1968	398.1%
1969	566.7%
1970	693.2%

The ultimate incurred loss is the total amount of all claims and legal expenses occurring in each year as actuarially projected to their final settlement. The results are expressed as percentages of the base year of 1960.

OUTLOOK FOR AN EVOLVING PROFESSION

"... today the definition of architecture, its concept by people everywhere, and its actual practice by the professional, is too limiting, too narrow in scope—only part of a larger picture." — Herbert H. Swinburne, FAIA

The next several decades will present the architectural profession with unparalleled opportunities. The entrepreneurial role in project development is only one of many promising fields beckoning architects with initiative and talent. Accompanying these unparalleled opportunities, however, are unparalleled dangers. For conservative architects who wish to remain in the familiar routine of the past, the outlook is ominous. A host of competitors—"megabuilders" practicing package building on a vast, city-building scale or specialist consultants chipping away at the architect's comprehensive services—is already beginning to exploit growing public dissatisfaction with the traditional building process. Conservative architects may survive, practicing their traditional design skills in a few quiet tributaries. But unless they augment their design skill, either by direct acquisition or through alliances with other professionals involved in the development process, they cannot hope to remain in contention for the big projects.

Expansion and diversification of architectural practice are part of an evolutionary reformation underway in the entire building industry. For the next decade or so, this reformation likely will manifest itself more in the industry's internal structure, management and methods of operation. Later, however, after the inevitable abatement of restrictive craft union rules, modification of archaic building codes, zoning ordinances and subdivision regulations, and the removal of other obstacles to technological progress and advanced planning concepts, the industry's products also will change.

As it currently is constituted, the building industry cannot handle the vast volume of construction needed by the year 2000. Already accounting for a roughly $100-billion volume in 1971, or 10 percent of the Gross National Product, the construction industry will assume ever greater predominance over the next few decades as the nation's largest industry. The industry's current practices are too costly, too slow and too undependable to meet these huge demands. Perennial shortages of skilled field craftsmen not only have run site costs sky high, they have limited the construction volume that physically can be put in place. An impending shortage of professional designers will reinforce the requirement for the industry's reformation.

THE ARCHITECT'S ANSWER In their response to rising client demands for faster, more economical, more dependable delivery of buildings, architects are bursting out of the restricted roles they have played over the past hundred years. Throughout the 20th century, architects have been restricted to a fairly narrow band of development services—generally the design and construction phases. But now they are applying their creative imaginations to the entire development process—conception, initiation, economic analysis, administration, financing—even the post-construction management of development projects.

The entire trend toward industrialized standardization is liberating the architect's creative imagination for more important work —the esthetic, social, environmental and even psychological aspects of design.

Associated with the architect's expanding role are a number of specialist subdisciplines—systems design, architectural programming, master planning, construction management, computer graphics, and research information storage and retrieval. Though some of these emerging subdisciplines are responses to technological changes, most are responses to client demands. Architectural programming has become ever more complex, as bureaucratic clients, both private and public, have begun replacing individual client owners. Increasing project size also complicates the programming process. Our tumultuous, heterogenous society is, in itself, a complicating factor. When Ictinus and Callicrates designed the Parthenon, the program focused on the apotheosis of the goddess Athena in a sacred precinct maintained by public money supporting the state religion. In modern America, public money seldom goes for public monuments. More goes for urban renewal projects, which often arouse social conflict. For conglomerate clients, comprising owner, user, governmental agencies and other concerned parties, an architectural programmer must be a master psychologist capable of eliciting the numerous facts, opinions and ideas needed to translate user needs into a physical solution.

PRESCRIPTION FOR SURVIVAL: TEAM! TEAM! TEAM! Whatever disagreement modern architects may have on other subjects, there is a virtual unanimity on one point: *the team approach to the development process is the wave of the future.* The philosophic axiom for survival states that the whole is greater than the sum of its parts. A modern architect is more like a symphony conductor, coordinating the efforts of many musicians, than a solo concert artist. The practice of testimonial architecture, created by an eccentric genius working alone, is virtually extinct.

One response to the new demands for cooperative team action on projects of ever-increasing size is a parallel growth in design firm size and composition. Judged by the recent past, the trend toward larger design firms offering programming, construction management, architectural, structural, mechanical and electrical design, and market and financial analysis

The architect's entry into development work accords with the AIA's aggressive new policies as active boat rockers.

will continue into the foreseeable future. The large firms' success in taking an increasing share of A-E business is attested by their proliferation. Between 1965 and 1970, the number of A-E firms with annual billings over $5 million more than doubled, according to Engineering News-Record's surveys of the top 500 design firms. This increase in large firms reflects client demands for accelerated project planning and administration.

The lesson for small architectural offices is clear. To survive the competition with large firms, they must form either alliances with other design firms or become specialist consultants offering services to fellow architects. As project size increases, alliances of all kinds—permanent, formal changes in organizational structure and temporary, ad-hoc joint-venture partnerships—are becoming more common. Mergers of architects, engineers, and planners to provide a range of services from design of regional sewer systems to interior design will doubtless increase in the future. Acquisitions, too, are increasing. Some large A-E firms recently have become subsidiaries of conglomerate corporations.

This last trend jeopardizes the architect's status as a professional, according to some architects. When an architectural firm goes public or becomes a corporate subsidiary, it inevitably must lose some degree of management control. Whether this loss in managerial autonomy entails a corresponding loss in professional status is merely another of the plentiful conundrums of modern life. The answer will obviously vary in individual cases. But if the answer is difficult, the problem can be stated fairly simply: How can the architect adapt for survival under proliferating competitive threats from

Houston's Fondren Square Shopping Center, a 33,000-square-foot development, was planned and designed by Starnes Group, Inc., in 1969. As its initial venture into development building, the firm took a limited partnership in the shopping center and nearby 14,000-square-foot office building. The firm has continued its development activity — usually as general partner or by taking full ownership under its own development corporation.

non-professionals and, at the same time, retain the special qualities that define his unique professional identity?

HOW TO EXPAND WITHOUT REALLY GROWING   For the small architectural practitioner fearful of compromising his professional status, development work can offer survival insurance. Despite the apparently inevitable monopoly on large A-E commissions by large firms, smaller projects will continue into the foreseeable future as a substantial, if diminished, proportion of total architectural work. By remaining alert to development opportunities in these projects, small architectural firms possibly can find their own route to survival, or even prosperity. By increasing their depth of services and by exploiting opportunities for equity investment, they should increase their annual income despite a diminished clientele.

The government's slowly expanding role in urban development offers promising opportunities for architects in master planning and other services far beyond the architect's conventional building design services. The New Communities Program, Title VII of the 1970 Housing and Urban Development Act, ultimately will accelerate the trend toward regional planning by providing federal grants and loan guarantees for new towns and planned-unit development projects. There is a new national commitment to the planning of our urban environment, replacing our traditional submission to the chaotic urban sprawl that radiates out from our central cities like the witless advance of a glacier. This national commitment to planning means more work for architects with the required skills.

Exemplifying the new opportunities opened by the emerging trend toward planned communities is the Woodlands development, a 17,000-acre new town to be located 35 freeway miles from downtown Houston, planned for an eventual population of 150,000 over a 20-year development period. The private developer, George Mitchell & Associates, plans to open entrepreneurial opportunities for the many architects engaged in designing the shopping centers, industrial parks, office buildings, research centers and other facilities planned for this huge new city.

The federal government's continuing use of turnkey construction also promises new development opportunities for architects. Apparently inspired by the success of turnkey public housing, the General Services Administration in late 1971 solicited proposals for its first turnkey project—for five federal office buildings in Illinois and Wisconsin. Among the respondents to the bid invitation were package builders and architect-contractor joint ventures.

NEW LIFE IN AN OLD PROFESSION   The architect's entry into development work accords with the AIA's aggressive new policies as active boat rockers instead of passive crewmen on the ship of state. At the 1971 AIA Convention, AIA President Robert Hastings sounded a call for individual political action. The architect, said Hastings, must "plunge actively into

Greenway Plaza — a new development project southwest of Houston — comprises an area equal in size to 95 downtown city blocks. Fifteen large office buildings will provide more Class A office space than was available in all of downtown Houston 25 years ago. Century Development Corporation also plans a medical complex, 400-room hotel, and 44 acres of underground parking. Architects for master planning are Lloyd, Morgan & Jones, of Houston. Century acquired the 127-acre site over a period of time and has leased back some of the land to its former residential owners until construction is ready to begin. The project is being built in two stages and target date for completion is 1985.

political life, enlist allies, swing votes, mobilize community action, and take political positions on issues that once were thought to be outside his rightful area of concern." Hastings denounced the current system that promotes sprawl, waste, neglect, and environmental destruction. AIA delegates enthusiastically supported a resolution calling for a national land-use policy.

For architects who remain in the professional mainstream as practicing designers, these aggressive new AIA policies promise remarkable new opportunities. They provide a collective organizational will energizing the emerging technological means for architects to strengthen their role as design leaders. The entire trend toward industrialized standardization is liberating the architect's creative imagination for more important work—the esthetic, social, environmental, and even psychological aspects of design. Future architects will study alternative designs for buildings that minimize energy consumption and consequent air pollution, housing projects that promote physical and psychological security, downtown complexes designed to encourage mass-transit and neutralize the hazardous, massive invasions by automobiles on city streets.

The team approach to the development process is the wave of the future.

Opportunities in this fast changing society never have been greater. Architects are uniquely equipped to determine not only *how* a project should be developed but *whether* it should be developed. The factors involved—judicious acquisition of land, a favorable mortgage commitment, and fast construction—are as important as design. In each of these areas, the architect can take a major—if not the leader's—position.

A wide range of development possibilities is involved in shaping the human environment. A new town, a research campus, a motel, a neighborhood swimming and tennis club—even the alteration of a back porch—can be development projects. They are some of the unlimited projects which the architect first can initiate and then create.

Whether to participate in entrepreneurial ventures is, of course, a matter of personal choice. Some architects may prefer to continue in their traditional roles, but many architects have begun earnest investigation of development team roles and the broad opportunities inherent in the process. Participation in the development team can do much to ensure the architect's continued active voice in shaping man's physical environment.

From a consumption-obsessed society in which the standard of living is measured solely by the Gross National Product, the United States has climbed to a higher, more sophisticated cultural level. Americans are beginning to appreciate quality as well as quantity. They are demanding environmental amenities and public beauty. The architect is uniquely qualified to deliver these goods, and the developer's role enhances his opportunity to do so.

GLOSSARY

Accelerated depreciation: Depreciation in which the deduction (from net return) starts at its highest annual value in the first year and steadily diminishes in the later years.

Amortization: Repayment of a loan in regular installments that continually reduce the principal, or outstanding portion of the unpaid loan balance.

Appraisal: A determination of property value, made by a qualified real estate appraiser or a person familiar with real estate values.

Assessed value: A valuation established by a public official (tax assessor) as a basis for local, county, and state real estate taxation.

Balloon payment: A large final payment terminating a debt.

Base: A term designating one of several alternative financing arrangements in a pro-forma projection.

Before-tax income: Gross income minus all expenses except income taxes.

Capitalization rate: Rate of return on net operating income considered acceptable for an investor and used to determine capitalized value. This rate should provide a return **on** as well as a return **of** capital.

Capitalized value: Estimated market value computed by dividing annual net income by capitalization rate. Capitalized value is thus appraised value that would yield a given annual net income at an assumed rate of return.

Cash flow: Money left from a project's gross income after all expenses—operating and debt-service—have been deducted. Some developers include depreciation and deferred tax calculation in computing cash flow.

Compensatory (or compensating) balance: A minimum non-interest-bearing bank account balance equal to a voluntary or agreed percentage (e.g., 20%) of an outstanding loan or credit line; it is maintained in order to furnish the lender with a greater effective return.

Constant: (see "Debt service constant")

Construction loan: A short-term loan enabling a developer to pay contractors' bills and other expenses incurred before and during the construction period (also known as interim loan).

Contingent interest: A lender's equity-sharing provision normally appended to a loan at a fixed interest rate and calling for a percentage of annual gross or net project income exceeding an agreed base amount. Contingent interest is paid to the lender in addition to fixed interest.

Debt capital: Money loaned at an agreed interest rate for a fixed term of years; distinguished from equity capital.

Debt service: A borrower's annual payment comprising repayment of principal plus payment of interest on the unpaid balance.

Debt service constant: A factor that, multiplied by the total loan amount, or total principal, yields the annual debt service payment (principal plus interest) required to amortize a loan.

Depreciation: A sum representing presumed loss in the value of a building or other real estate improvement, resulting from physical wear and economic obsolescence, and deducted annually from net income to arrive at taxable income.

Development loan: A short-term loan, advanced before a construction loan, used by developers to acquire land and install basic utilities—roads, sewers, water supply systems, etc.

Discounted loan: A loan whose nominal value is reduced by a certain percentage (or "points"). On a $1000 loan discounted 3 points, the borrower actually receives $970. He repays debt service on $1000.

Development process: The process through which development projects are conceived, initiated, analyzed, financed, designed, built, and managed.

Economic design: A general term denoting the entire scope of project economics—market and feasibility studies, plus owner's investment analysis.

Equity: (1) Ownership and its concomitant right to all or part of a project's profits; (2) the owner's share of total value.

Equity capital: Money invested by owners or others who share in profits; distinguished from debt capital.

Feasibility study: An analysis of a proposed project's economic prospects, used as the basis for the developer's decision to build.

Financing: The process through which a developer obtains the capital he needs to purchase land, design, build and even manage a development project, either through loans—e.g., construction (interim) and mortgage (permanent) loans—through equity sharing or through more complex techniques—e.g., sale-leaseback.

Finder's fee: A fee usually in "points" paid to a banker, broker or other middleman who located debt or equity capital sources for a developer.

Foreclosure: A legal procedure in which property mortgaged as security for a loan is sold to pay a defaulting borrower's debt.

Front money: Equity capital used for expenses incurred in initiating a development project—e.g., land-purchase option, legal fees, administration costs, etc.

Gap financing: A loan required by a developer to bridge the gap—i.e., to make up a deficiency between the amount of mortgage loan due on project completion and the expenses incurred during construction.

Gross income: Total project income before any expenses are deducted.

Highest-and-best-use study: A comparative analysis of two or more alternative projects already judged economically feasible to identify the most profitable use.

Interim loan: Synonymous with construction loan.

Joint venture: An ad-hoc partnership formed for a limited, specific purpose by investors in a development project. The joint-venture agreement establishes the partners' duties in the development process and specifies how the ownership and profits are divided.

Land-sale-leaseback: A sale-leaseback negotiated on land only, enabling a developer to build on a leasehold.

Lease: A contract between owner and tenant, setting forth rental rates, term of occupancy, and other conditions.

Leasehold: The estate—usually land, or land plus building—held by the lessee under a lease.

Lessee: The tenant who leases property.

Lessor: An owner who leases property to a lessee tenant.

Leveraging: The technique, usually through borrowing, of maximizing an investment's profit-equity ratio.

Lien: A legal claim on a property for payment of a debt secured by the property or some other financial obligation—e.g., mortgage, taxes, unpaid repair or construction bill.

Loan fee: The charge made for negotiating a loan, in addition to interest (see Finder's fee for payment of); sometimes used with reference to an additional fee paid directly to a lender either for a commitment or at the time advances are made.

Market study: A projection of future demand for a specific type of project, usually with a recommendation for volume of space to be sold or rented and sale or rental price.

Market value: The price at which a property could be sold on the open market, with buyer and seller free from abnormal pressures.

Marketability study: A more narrowly focused study than a market study, restricted to recommendations for volume of space and price.

Mortgage: A lien on land, buildings, machinery, or other property, pledged by a borrower as security to a lender (sometimes called a deed of trust).

Mortgage commitment: A legal contract between a borrower and a lender to advance a mortgage loan when the borrower meets certain conditions—e.g., completing the project, acceptance by the lender's agent, etc.

Mortgagee: A lender who advances a mortgage loan.

Mortgagor: A borrower who gives a mortgage on his property.

Net net: Net project income after deducting insurance and maintenance expenses, but not real estate taxes.

Net net net: Net project income after deduction of insurance premiums, maintenance expenses, *and* real estate taxes.

Net operating income: The remainder left after total operating expenses (exclusive of interest payments) are deducted from gross income (same as "Net net net").

Net return: The remainder left after total operating expenses *and* interest payment are deducted from gross income.

Permanent loan: A long-term mortgage loan, distinguished from a short-term interim or construction loan.

Points: The percentage deduction from the nominal amount of a discounted loan, often charged as a finder's fee. On a $1000 loan discounted 2 points, the borrower receives $980 (1 point$=$1%).

Prepayment penalty: A penalty payment, in addition to interest and principal, sometimes levied for repayment of a loan before it falls due.

Prepayment privilege: The condition under which a borrower can repay a loan before it falls due with or without incurring a prepayment penalty depending upon the agreement.

Principal: The amount of debt, exclusive of accrued interest, remaining on a loan. Before any principal has been repaid, the total loaned amount is the principal.

Pro forma: Latin for *according to form;* a projection of anticipated annual income, expenses, and cash flow from an investment enterprise, indicating the form in which the data should be presented.

Project administrator: The client's representative in administering a construction project (sometimes known as project manager).

Purchase-money mortgage: A mortgage given to the seller as all or part of the purchase consideration in exchange for property, most commonly used in land purchases, with prior rights over any subsequent lien; unless made subject to subordination.

Rental attainment provision: A mortgage commitment clause requiring a minimum occupancy level in a project before the full amount of the mortgage is advanced.

Sale-buyback: A financing arrangement in which the developer sells a property to an investor and then buys it back on a long-term sale contract; this is sometimes called an installment-sale contract.

Sale-leaseback: A financing arrangement in which a developer sells all or part of a project and then leases the portion he has sold.

Second mortgage: A mortgage that has secondary rights to the first mortgage—i.e., the proceeds from a foreclosure sale must pay the first mortgagee before any funds can go to repay the second mortgagee.

Secondary financing: A loan secured by a second mortgage on a property; sometimes used to refer to any financing technique other than equity and first-mortgage debt.

Stabilized: A term used in pro forma projections to denote averaging of perennial expenses or income that may change over the life of the project or the term of the mortgage loan.

Standby commitment: A mortgage-loan commitment issued by specialist lenders and accepted by another lender as a basis for advancing a construction loan.

Standby fee: A 1- or 2-percent good faith fee submitted, along with the application for the permanent loan, to the lender.

Straight-line depreciation: Depreciation deduction in constant annual amount.

Subordination clause: A clause in a junior lien acknowledging the prior claim of a higher loan—as in a second-mortgage loan contract legally acknowledging the prior claim of the first mortgagee; also describes an agreement contained in purchase-money mortgages for land by which the purchase-money-mortgage can be subordinated to a first mortgage to finance bona fide improvements.

INDEX